# Design Projects and Concepts: Architectural and Specialty Design Works to Promote the Independence for People with Disabilities

## By

## Maureen Gaynor

A Companion Book to
*What If Nobody Finds Out Who I Am*
By Maureen Gaynor

# Adaptive Bookstand/Headboard
## For Futon on Floor

This headboard measures 54" wide by 8" high by
8" deep.  The slanted bookstand portion measures 24" wide, flanked by 15" flat sections.

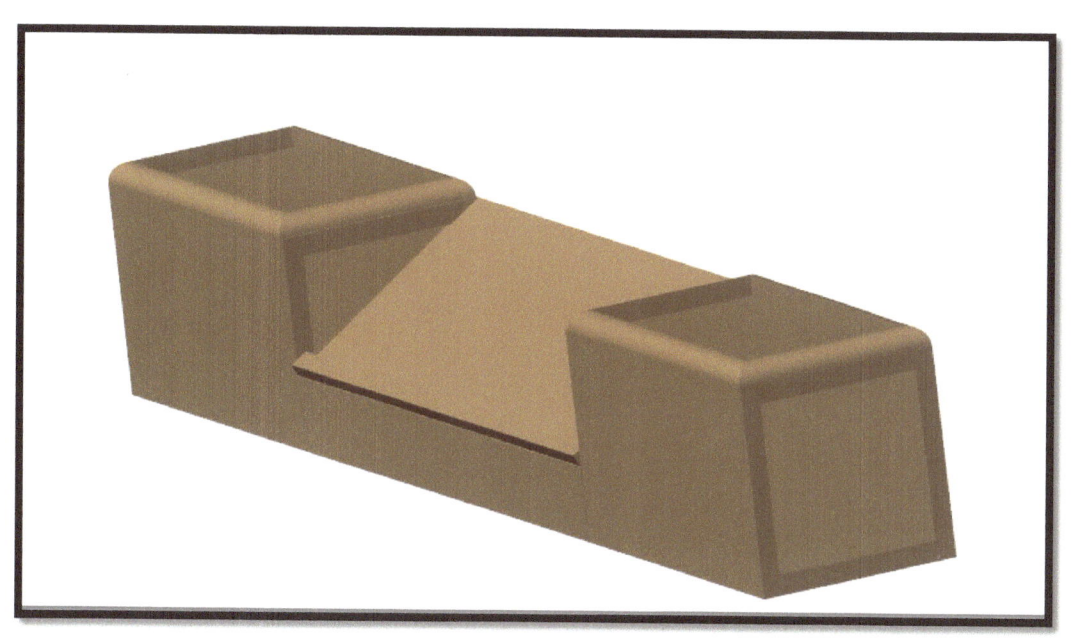

Rounded molding is placed around the flat sections so objects won't fall off.

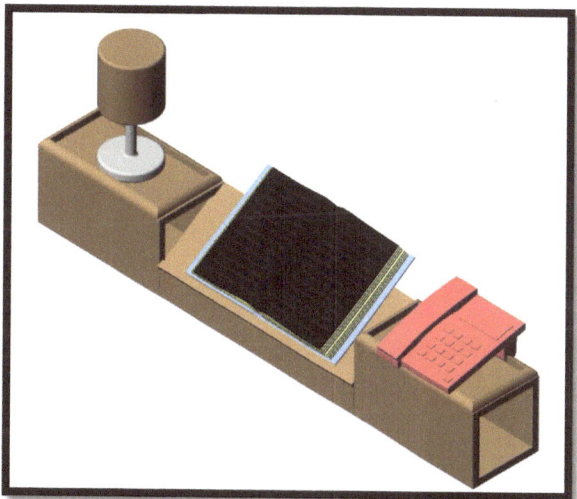

# Adaptive Cashier Ramp
## (Designed by Andrew Galloway)

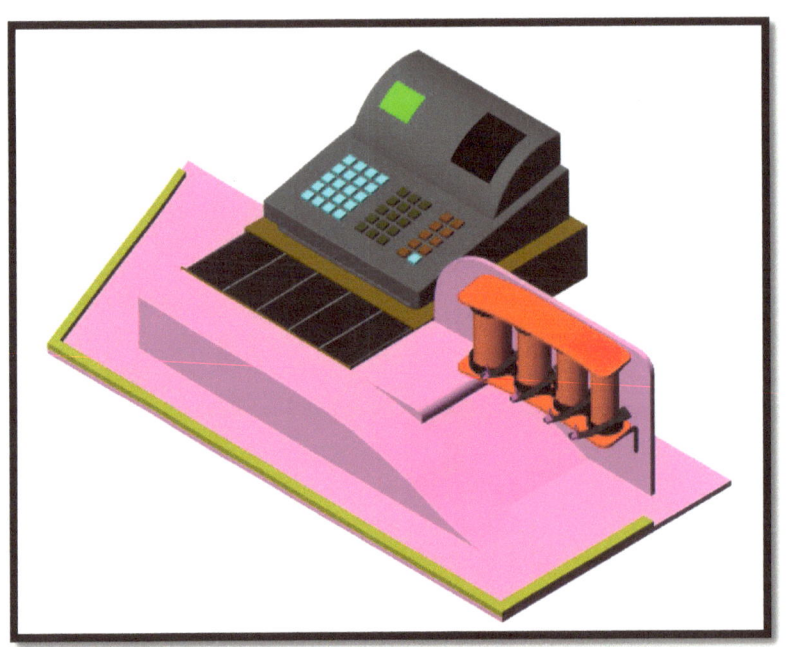

Close-Up of Coin Dispenser, attached to a vertical panel by brackets.

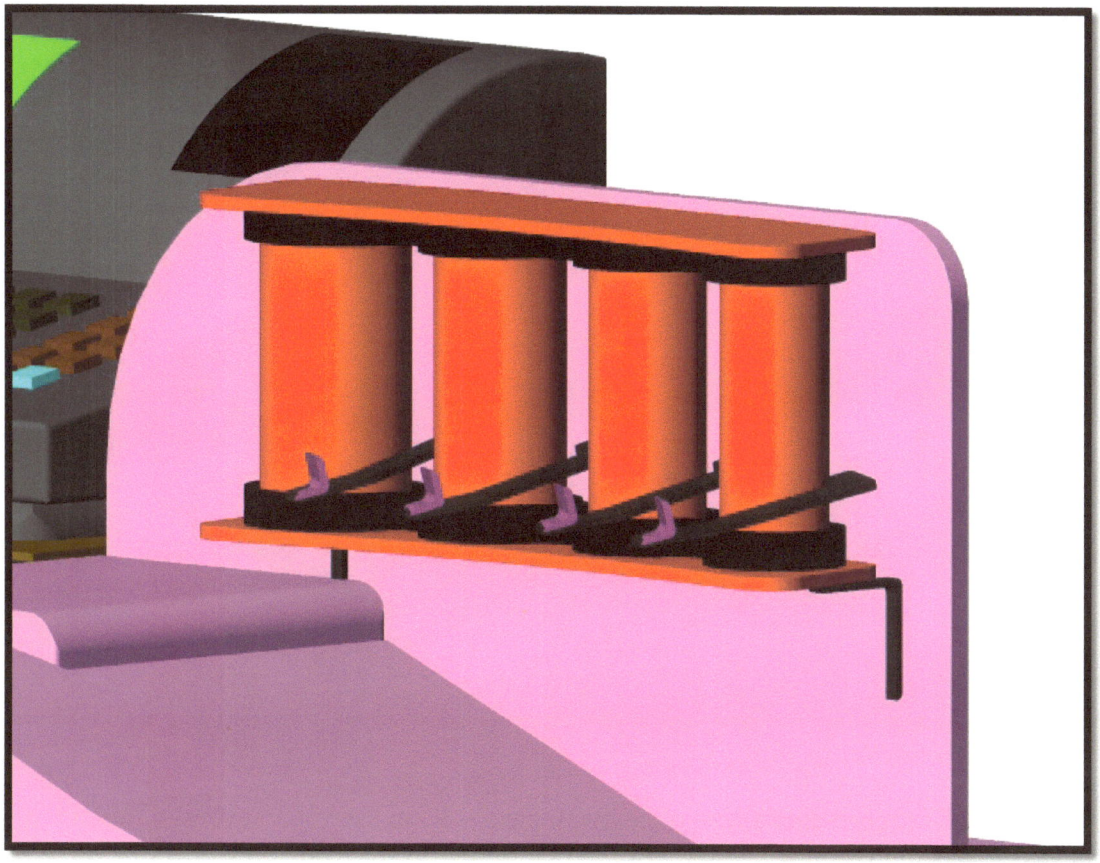

Note: L-Shaped metal objects are attached to the levers of the coin dispenser to easily press down with headpointer or finger.

# The 1980's Setup for Brushing Teeth

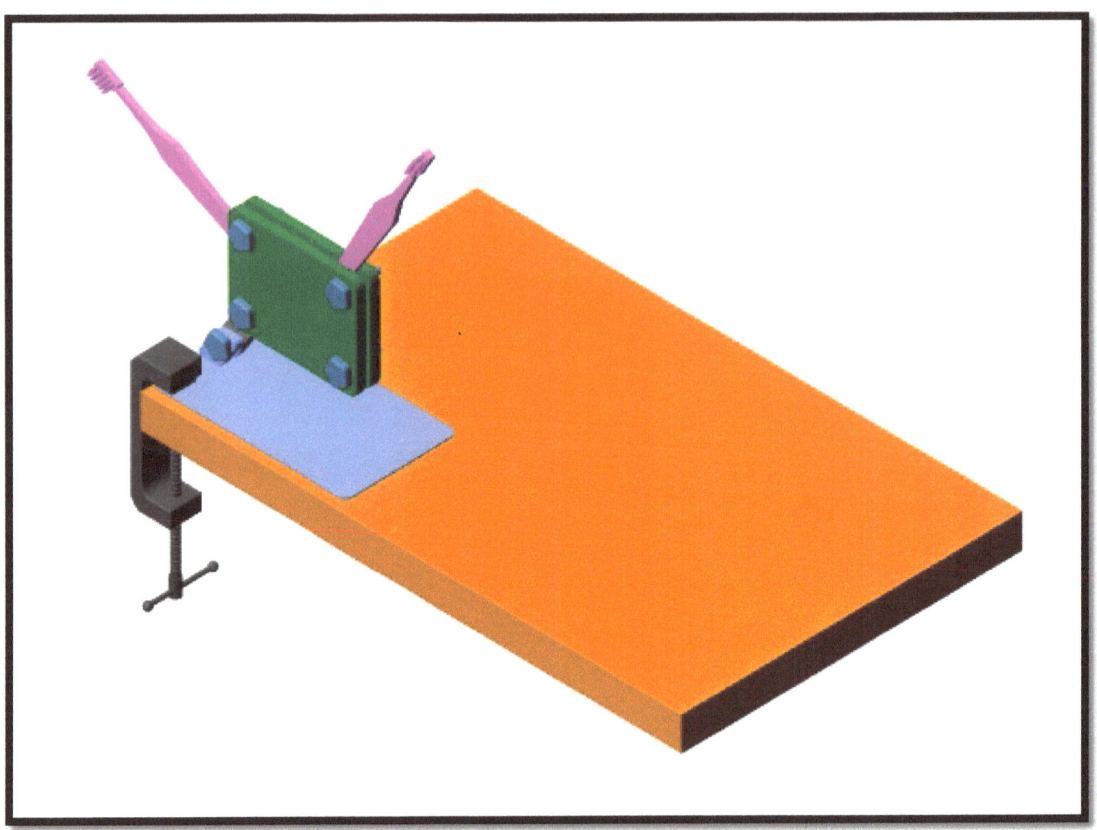

The orange surface represents any surface available to clamp the apparatus used to hold the two toothbrushes at different angles. The use of the two toothbrushes made the apparatus more efficient since the apparatus couldn't rotate the full 360°.

The two metal plates, represented in green, secured the two toothbrushes in place. Four individual wing nut screws were easily tightened or loosened when the toothbrushes needed to be cleaned and repositioned.

Even though this toothbrush apparatus looks very simple, and by today's standards, antiquated, I thought it worked very well and I enjoyed the independence of using it.

The toothbrush apparatus rotated in a forward position.

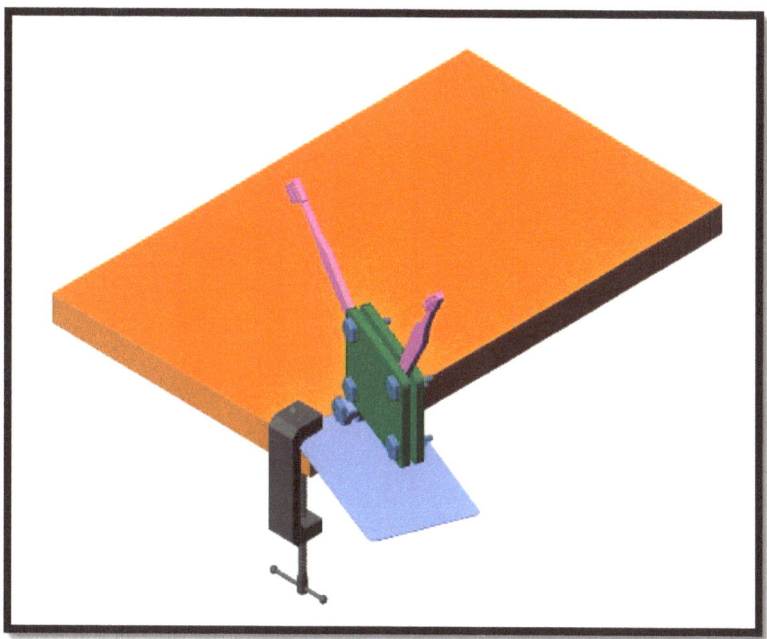

The toothbrush apparatus rotated in a backward position.

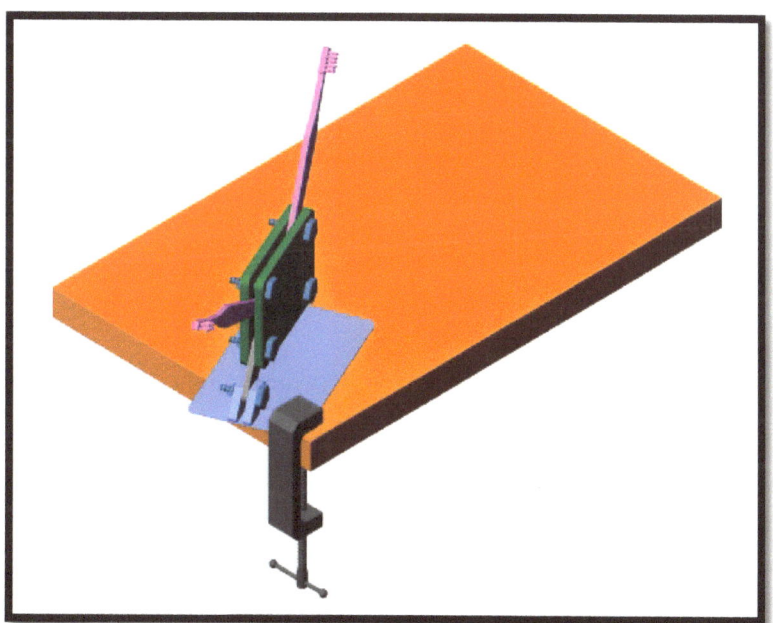

# Adaptive Toothbrush Holder: New Concept

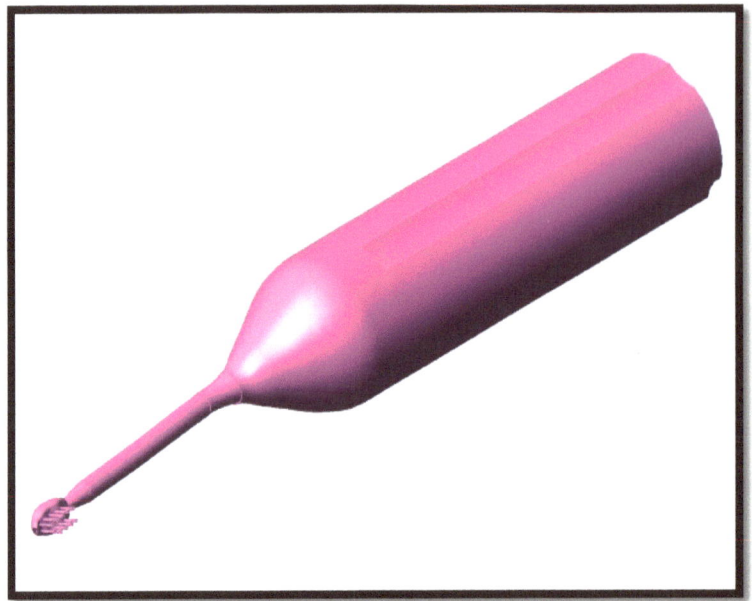

An average electric toothbrush

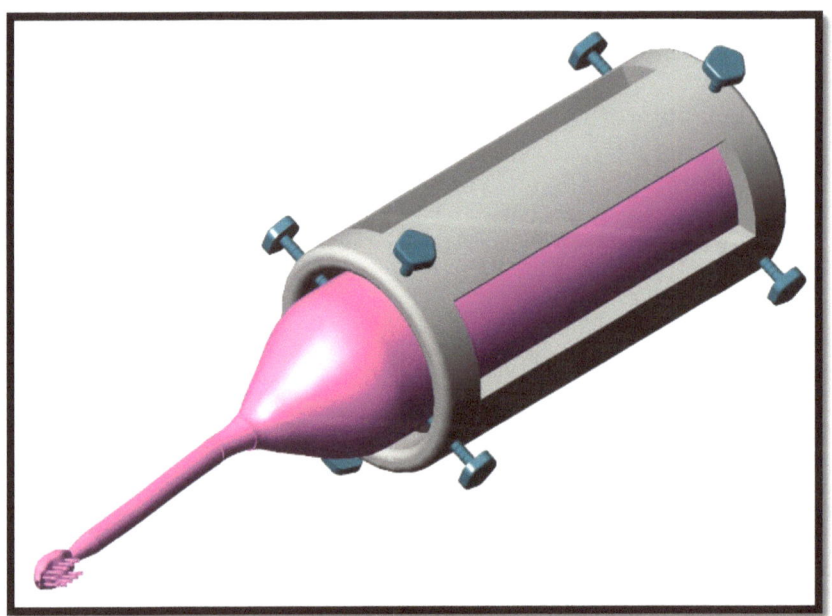

Figure 1

Figure 1: I created a open cylinder, with a diameter of 2" and height of 4.5". The cylinder has 8 separate tightening screws to secure toothbrush in position.

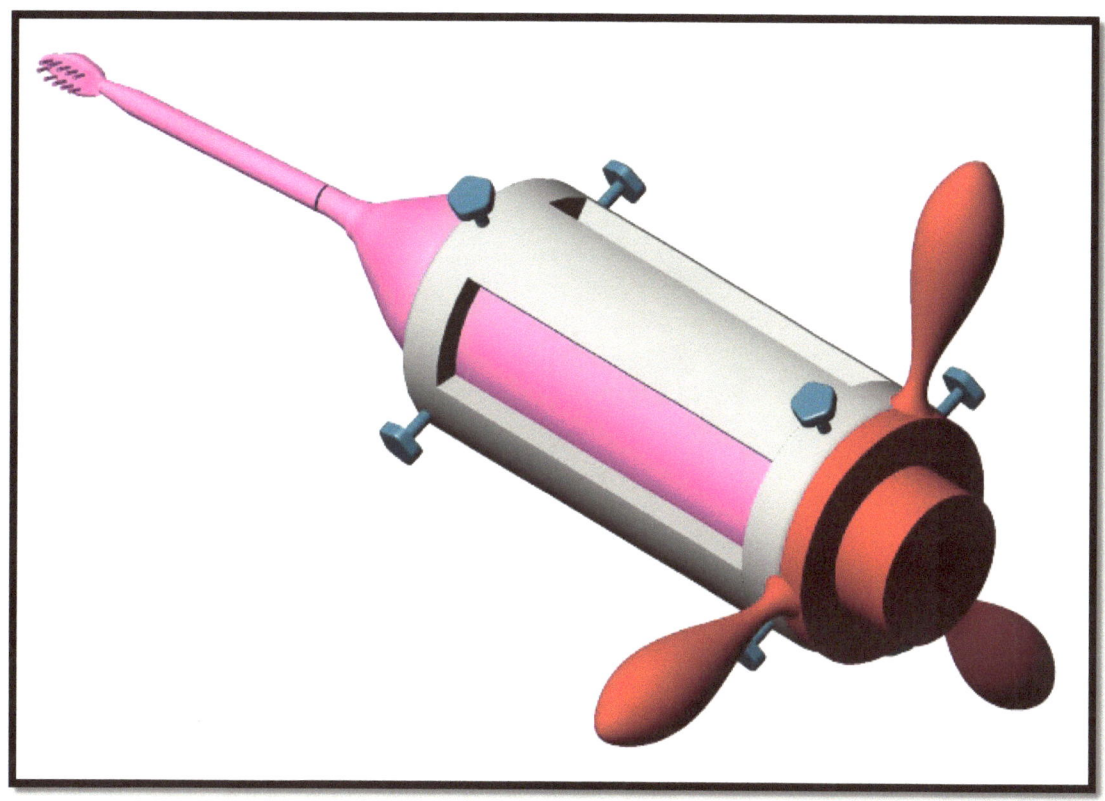

Figure 2

Figure 2: I created and attached a 360° rotating gear with 3 projecting pegs for the user to rotate the toothbrush easily. The projecting pegs, rounded and contoured to prevent scrapes, extend outward 3" for easy turning by a single hand, finger or any possible way the user can rotate the cylinder.

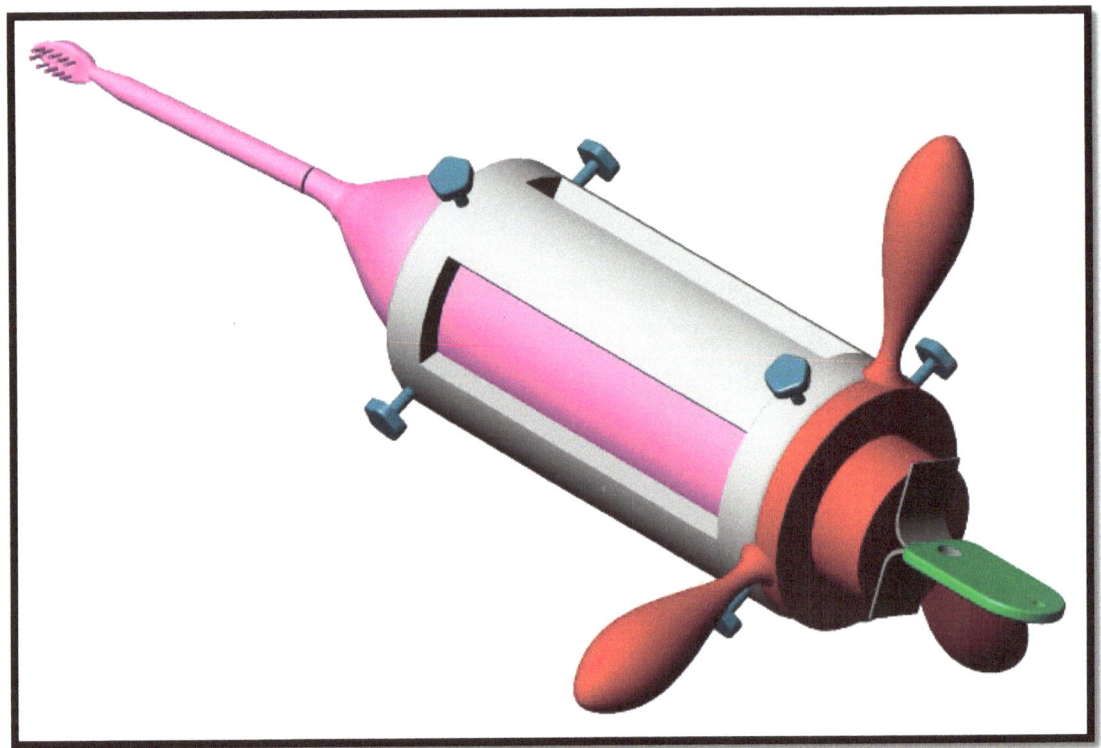

Figure 3

Figure 3:  Attached to the rotating gear is a metal hinge.  The hinge, green, has two separate holes. The smaller hole, 1/16$^{th}$" dia., positioned nearest to the rounded bottom edge is the rotating center, which the most of the toothbrush mechanism will swing forward 90°.  The larger hole, .25" dia., positioned nearest to the L-Shaped metal brackets, is the hole securing the cotter pin when the toothbrush mechanism is in the 90° forward position.

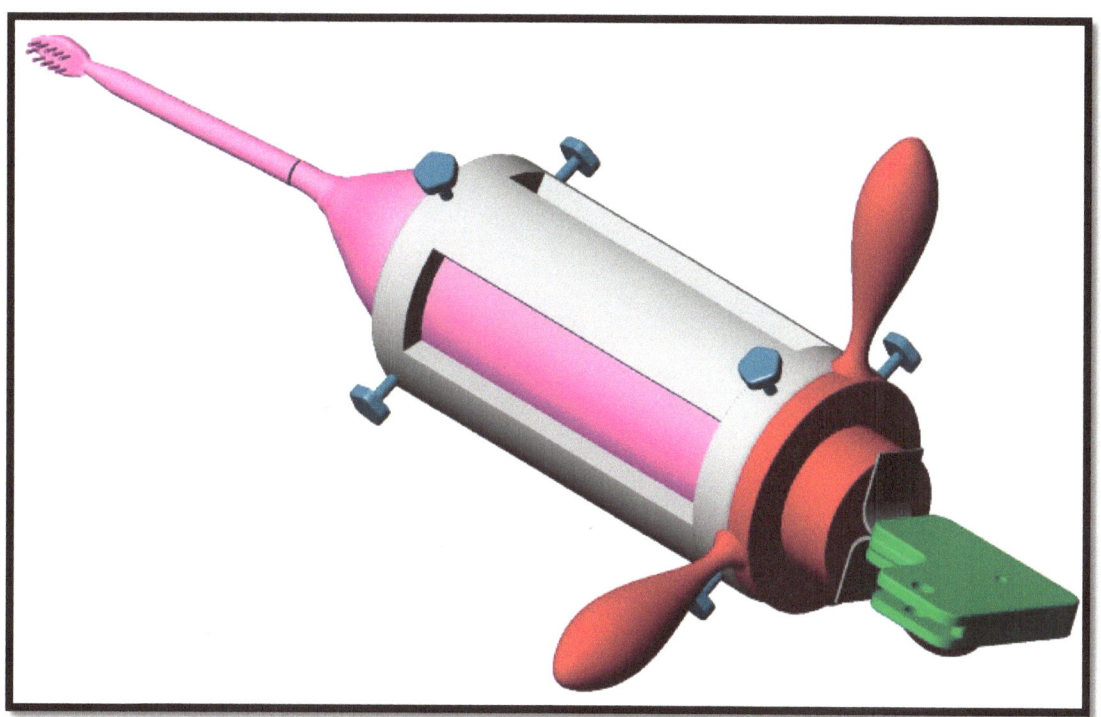

Figure 4

Figure 4: This illustration shows the outer body of the hinge, where the toothbrush mechanism will only be able to rotate forward 90°. Note the outer body of the hinge also two separate holes. The 1/16th dia. hole is directly in line with the 1/16th,, dia. hole in Figure 4, creating a pivot point for the toothbrush mechanism to rotate forward 90°. The ,25" dia. holes for the cotter pin on the outer body are positioned 90° counter clockwise where the cotter pin will lock the toothbrush mechanism in place.

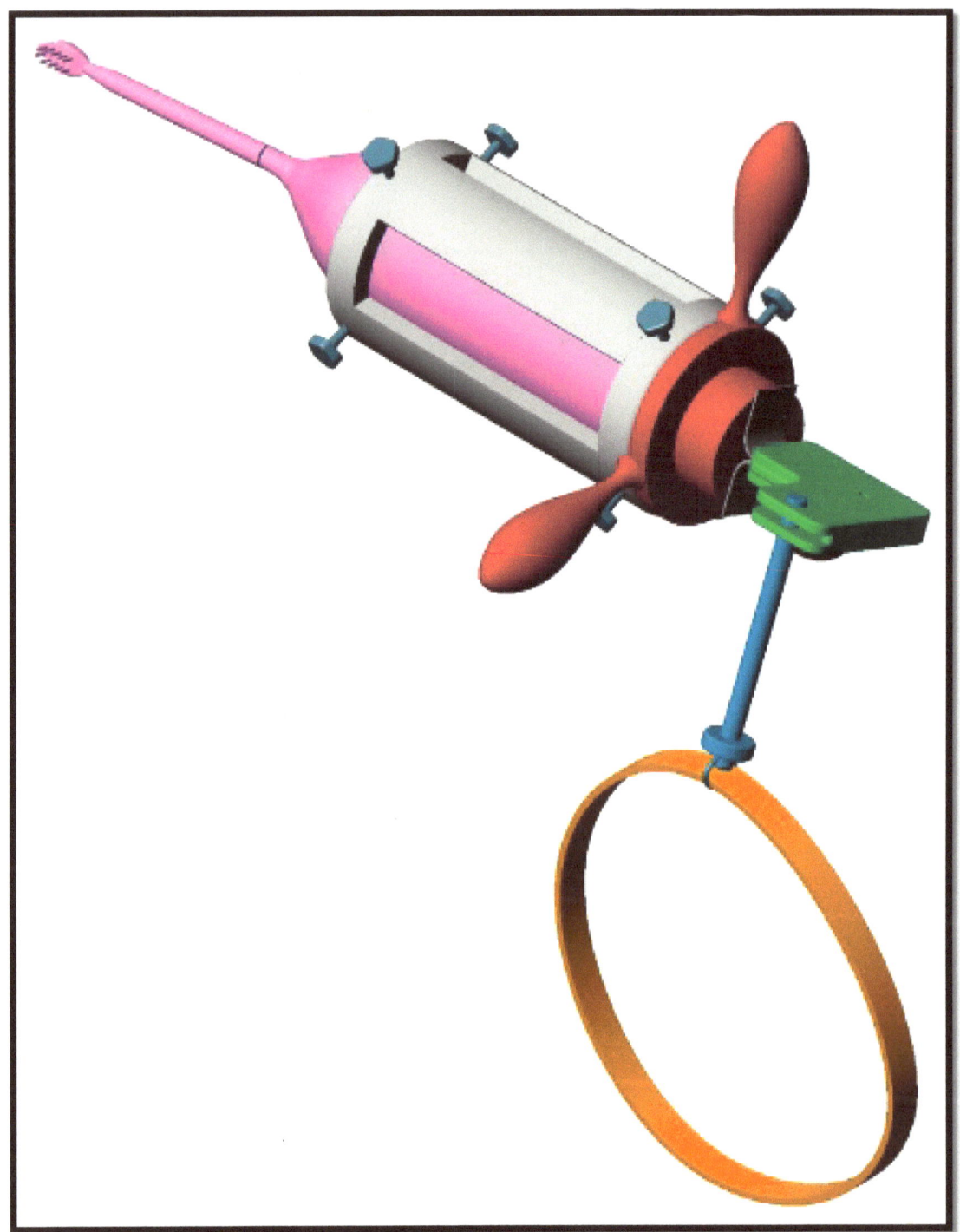

Figure 5

Figure 5: This illustration shows the cotter pin in position to lock the toothbrush mechanism in its forward 90° position. A fabric loop is attached to the end of the cotter for the user can pull down on the cotter pin easily to release the toothbrush mechanism out of its 90° position, as shown.

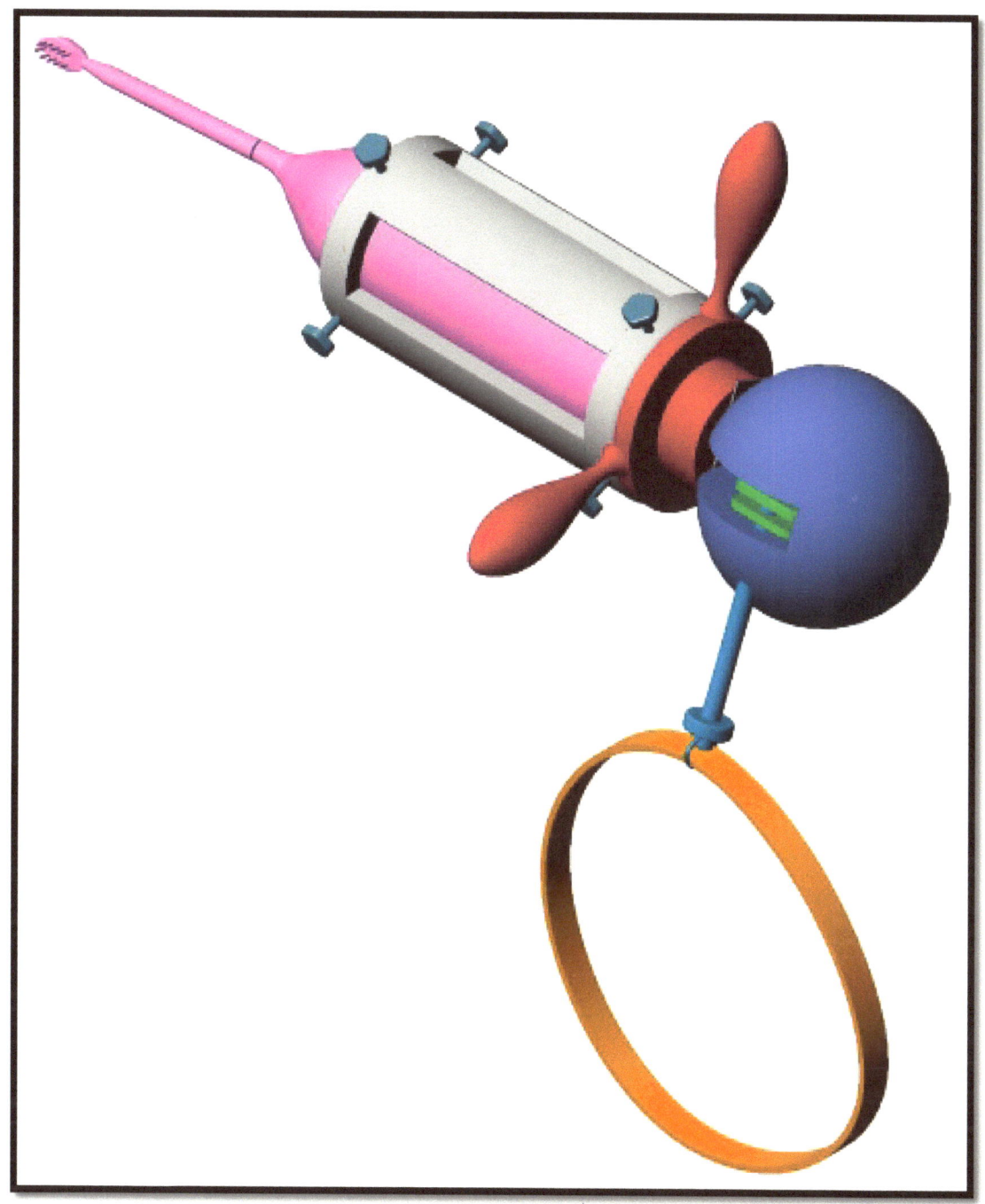

Figure 6

Figure 6: This illustration shows the metal hinge structures enclosed in a hard, plastic cover for protection.

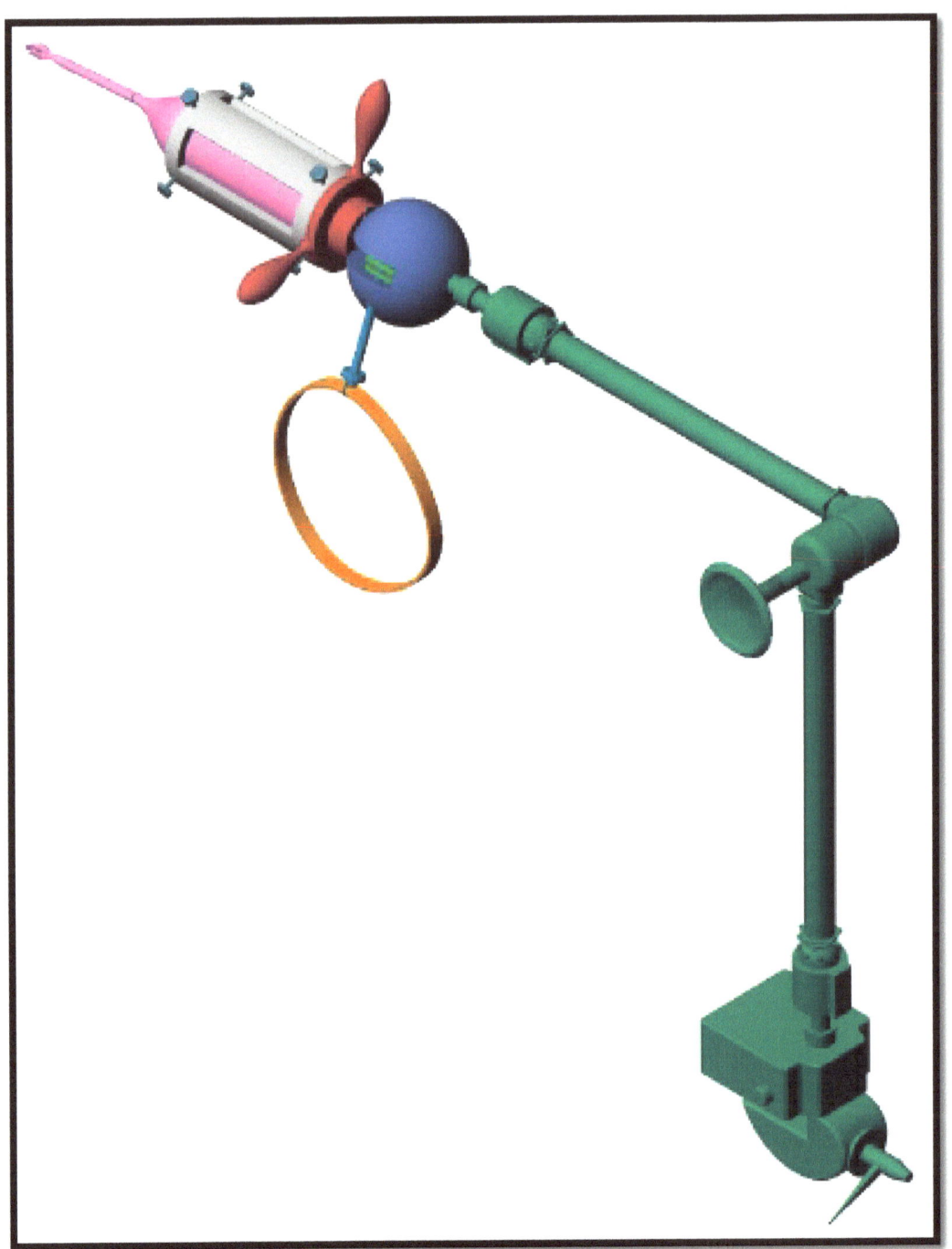

Figure 7

Figure 7: This illustration shows the toothbrush mechanism fixed on an universal mount, which can attach to a surface top and can be adjusted to any position to satisfy the individual user.

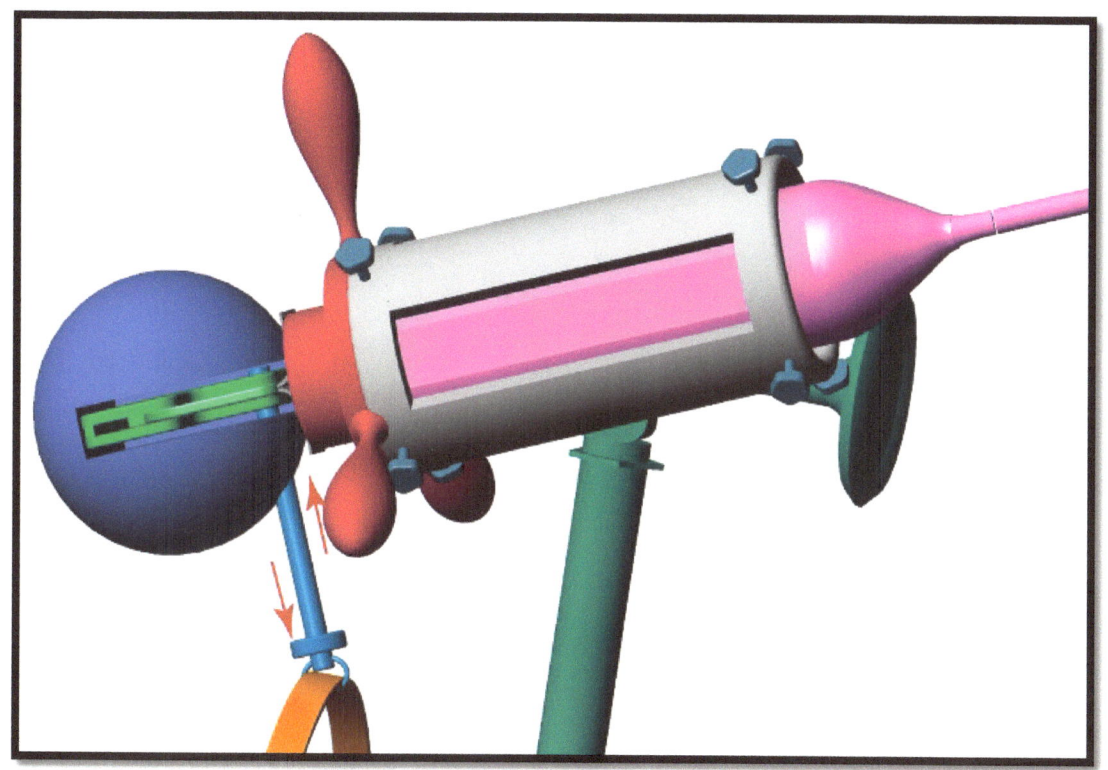

Figure 8

Figure 8:  This illustration shows the toothbrush mechanism locked into the forward 90° position. The cotter pin springs upward to lock the toothbrush mechanism through the two outer .25" dia. holes on the outer body of the hinge.  Once the cotter pin is pulled downward, the toothbrush mechanism returns to its linear position (Figure 7)

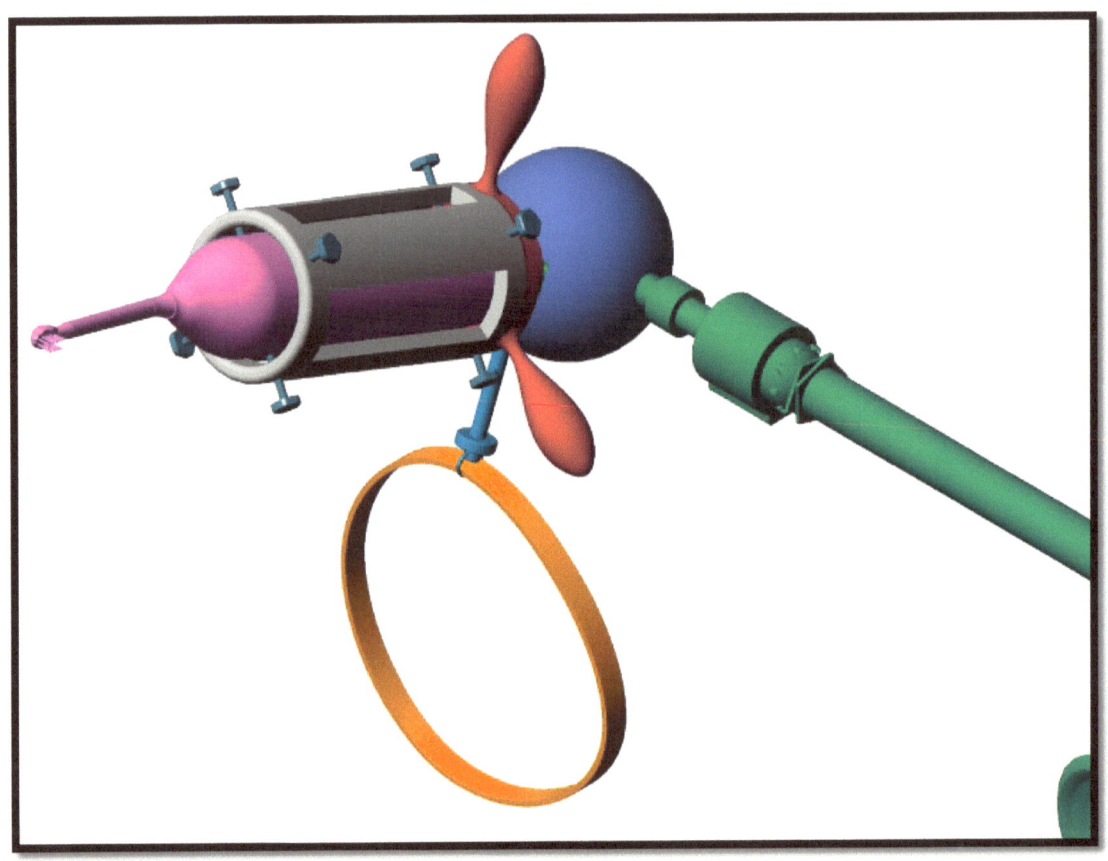

Figure 9

Figure 9: This illustration shows the toothbrush mechanism in its forward 90° position from a different perspective.

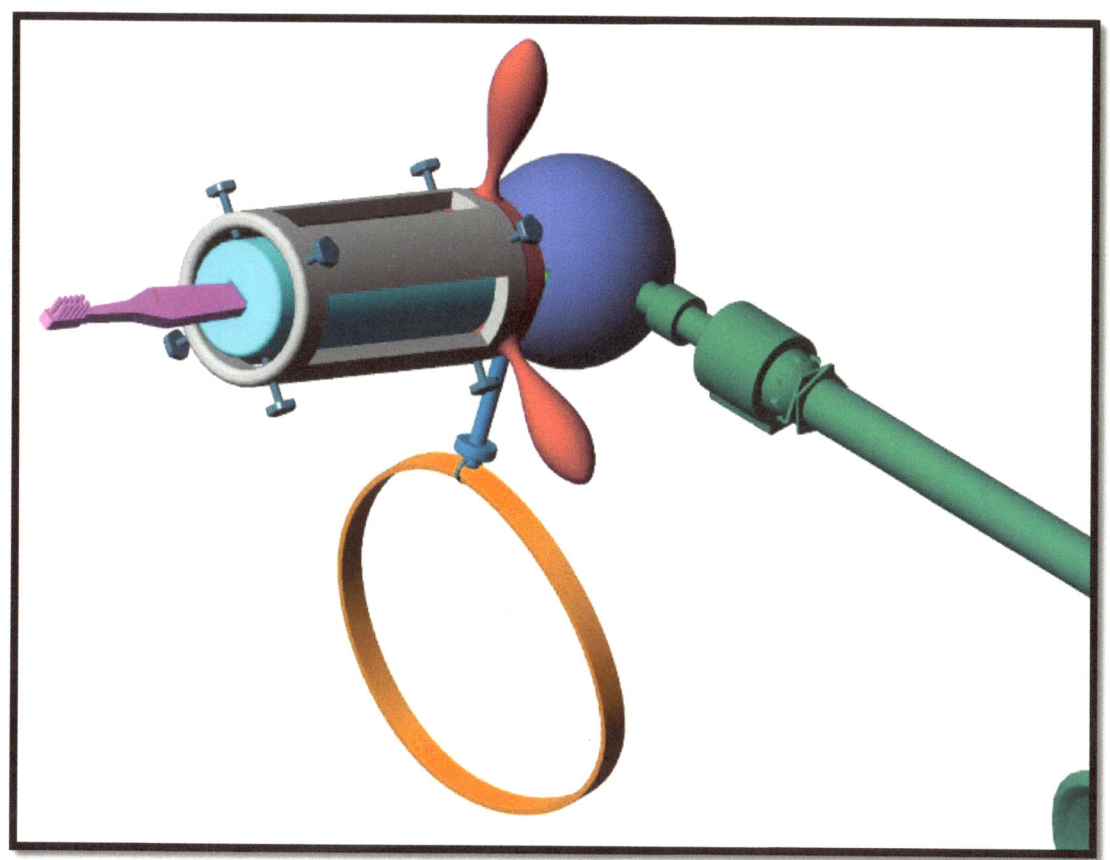

Figure 10

Figure 10:  This illustration shows an manual toothbrush insert placed in the toothbrush shaft if the user chooses to use a manual toothbrush.  Like the electric toothbrush, the eight separate tightening screws can secure the insert.

# The Adaptive Lavatory

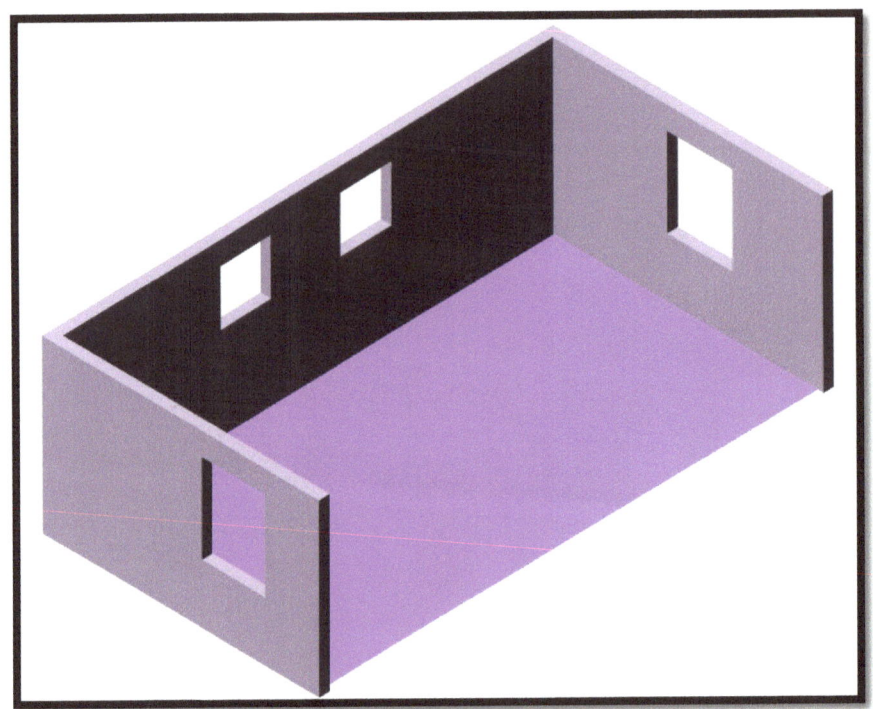

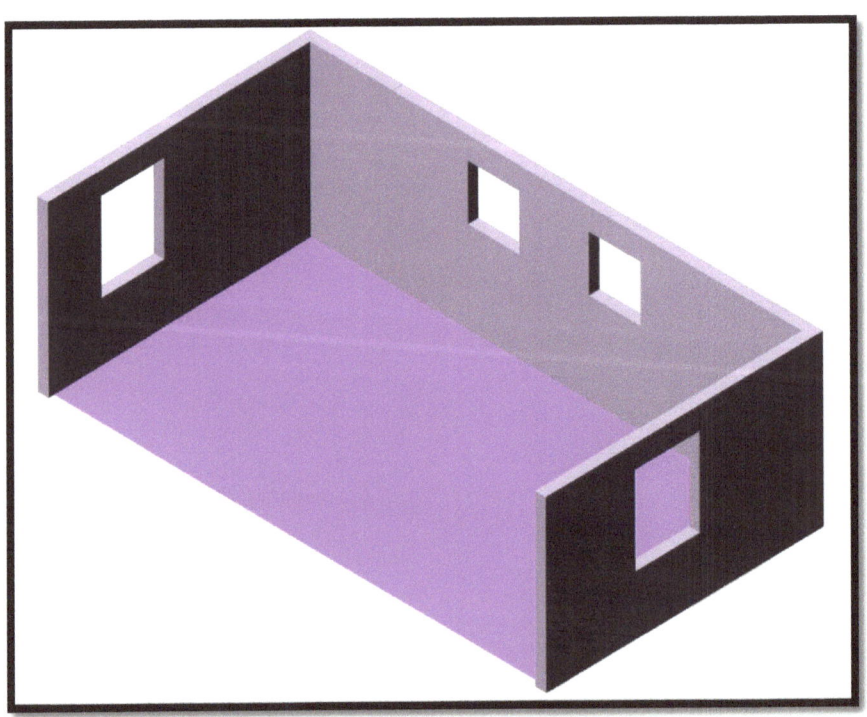

Bedroom measures 24' x 12'

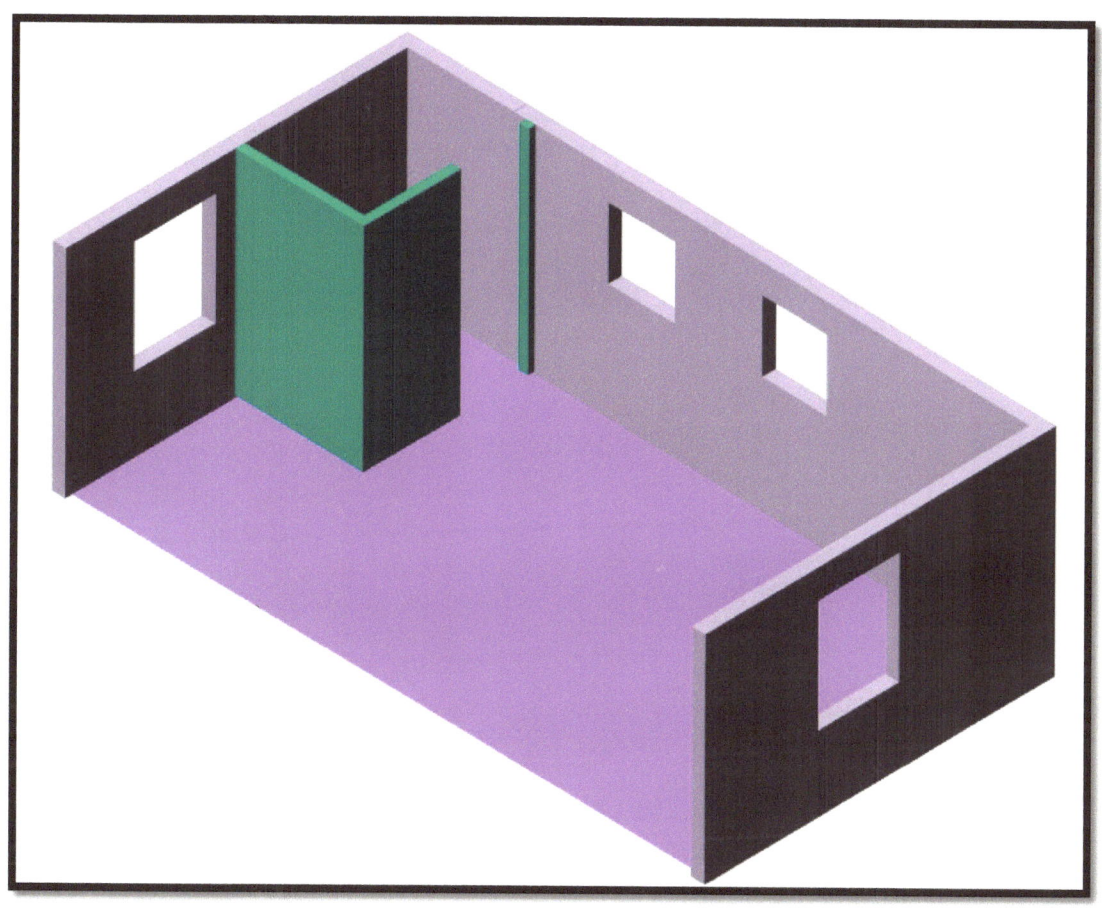

Figure 1

Figure 1: New walls were erected to construct the new, adaptive lavatory. The lavatory measures approximately 6'-4" x 4'-3".

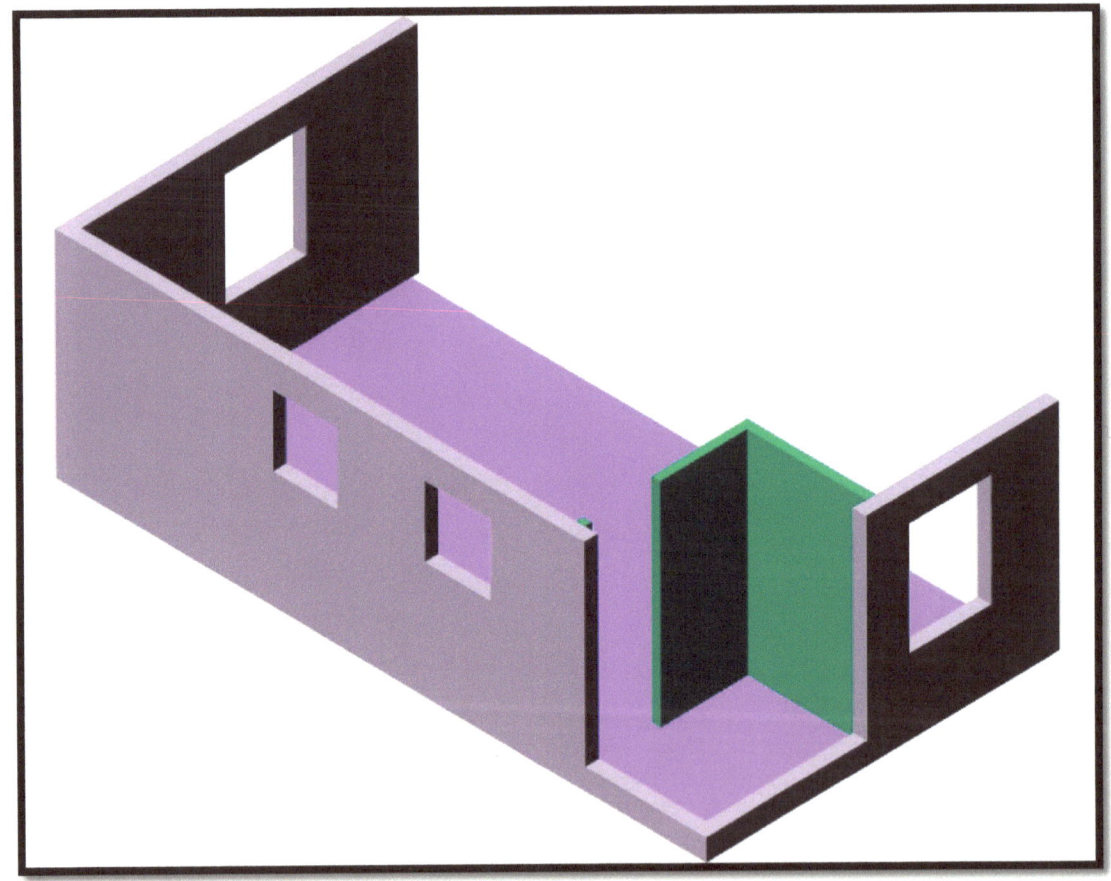

Figure 2

Figure 2: This illustration shows a cut-away view of the new walls of the lavatory from the opposite perspective angle. With the wall cut away, you will see exactly how I designed the lavatory.

Figure 3

Figure 3:  This illustration shows the installation of a standard height toilet.  Underneath the toilet, a water-resistant metal pan with a drain in case of leaks or overflows.

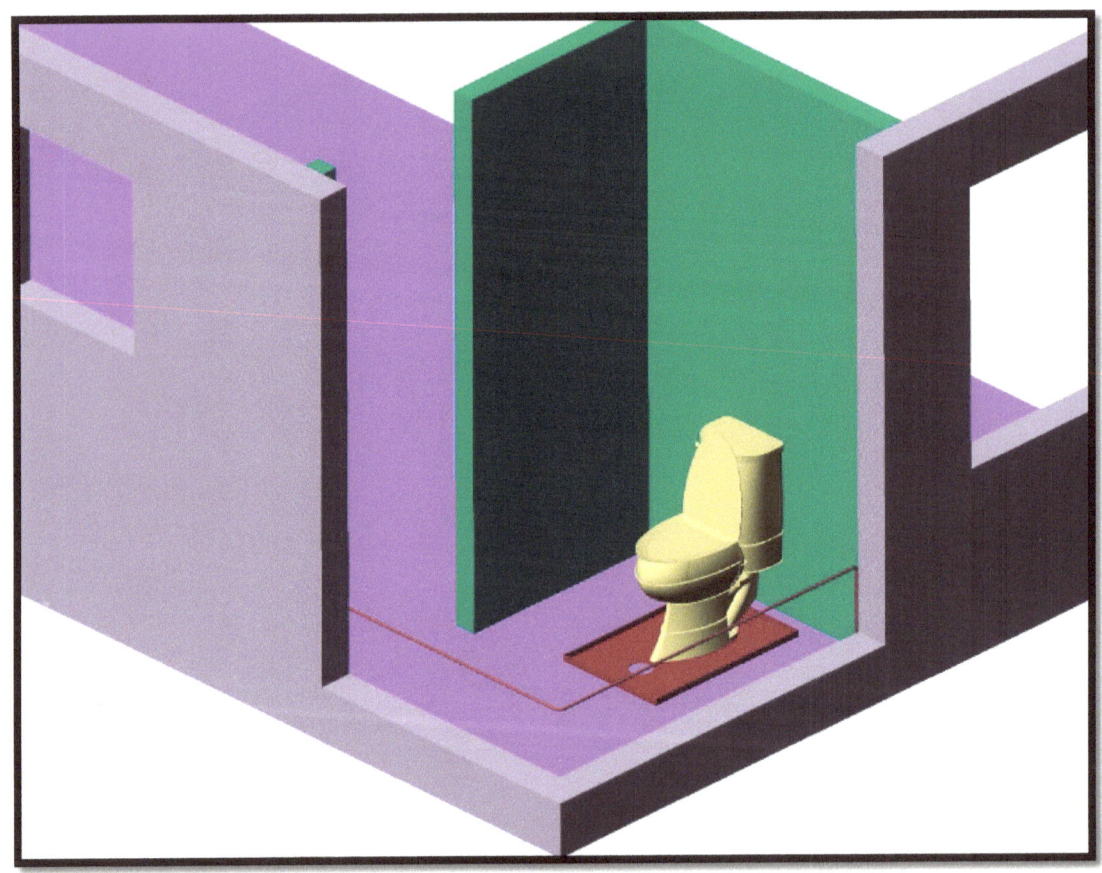

Figure 4

Figure 4:  Along the outer walls of the new lavatory, piping was installed 17" off the finished floor to accept the baseboard heating radiators.

Figure 5

Figure 5: Around the exterior of the lavatory, a ledger board, 2" x 6", was nailed into the existing studs. The top of the ledger board was 14" off of the finished floor.

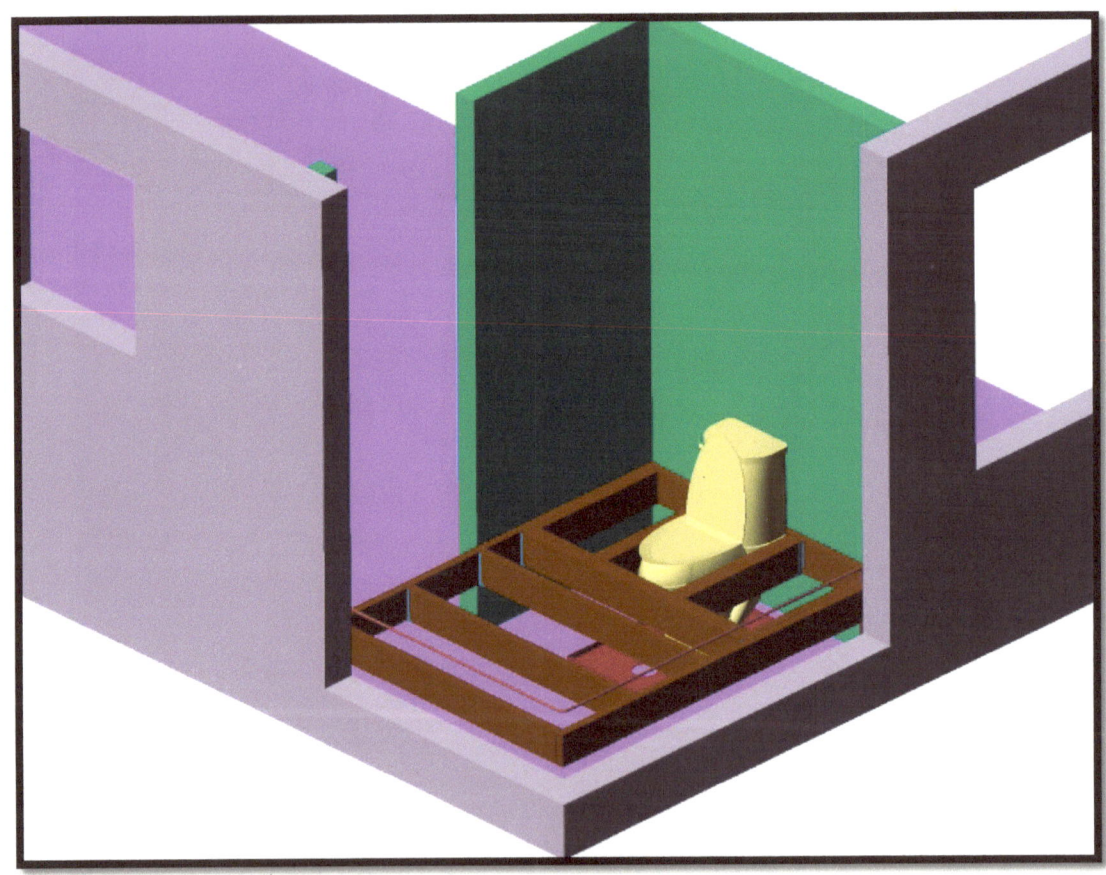

Figure 6

Figure 6:  Inner joists were positioned in place, very much in the fashion of a rough opening would be constructed around a fireplace or a staircase.

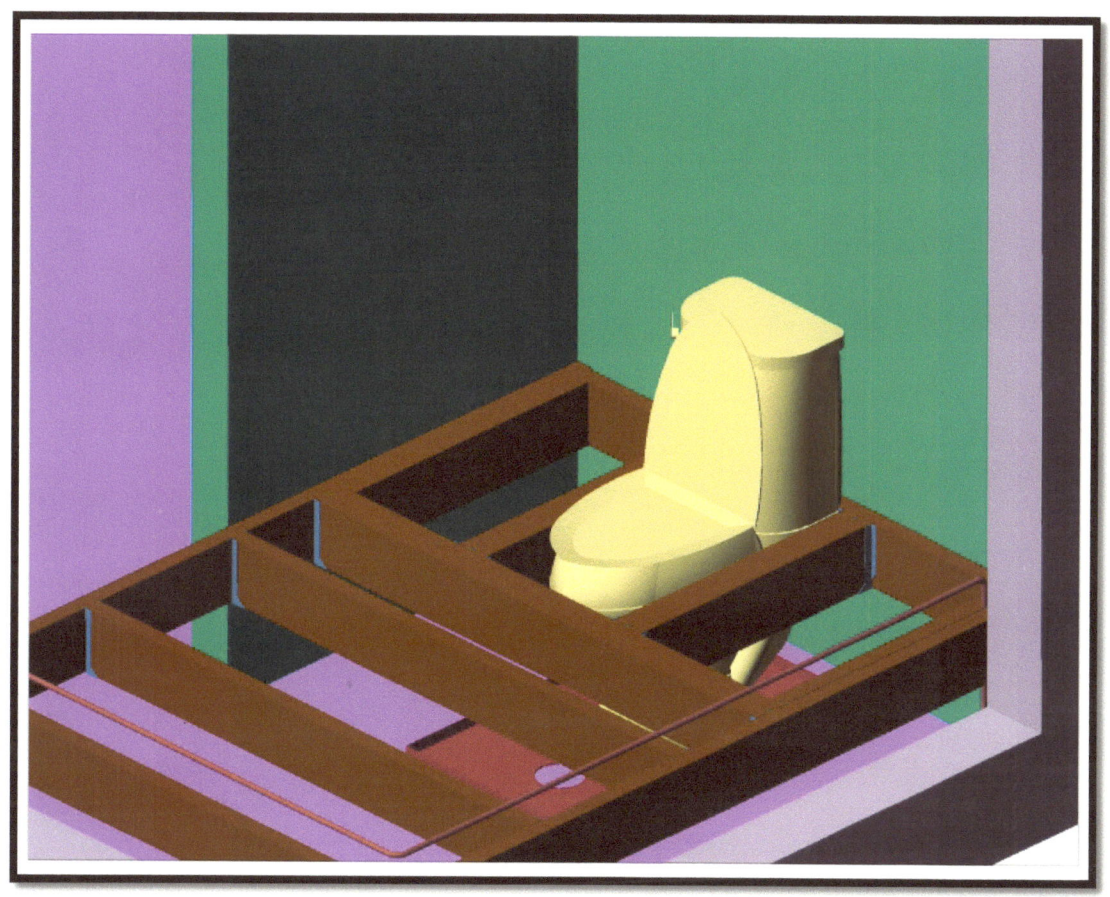

Figure 7

Figure 7:  In this close-up illustration shows the use of joist hangers for the inner 2" x 6" joists.  The inner joists are not nailed into the joist hangers to give easy access to the lavatory when the toilet needs to be replaced.

Figure 8

Figure 8: This illustration shows the installation of the baseboard heating in the new lavatory 15" off the finish floor.

Figure 9

Figure 9:  Three separate pieces of ½" thick plywood was cut and fitted onto the joists.  Again, the three pieces of plywood are not nailed down to give easy access to the toilet.

Figure 10

Figure 10: Over the plywood, hard foam covered in vinyl material was installed to protect the user's knees and hands from crawling on hard surfaces. The vinyl material is easily cleaned rather than soft fabric.

Figure 11

Figure 11: This illustration shows the framework of the 5" steps used to build up the multi-level platform up to the new lavatory, and to the bed. The width of the platform is 4' wide, and the center platform has a depth of 24".

Figure 12

Figure 12:  The multi-level platform was sheathed in ½" thick plywood.

Figure 13

Figure 13:  The multi-level platform was covered with foam and carpeting, with all the edges and corners rounded and padded to prevent injury.

Figure 14

Figure 14:  Beside the multi-level platform, a platform and a headboard was built for a mattress.

Figure 15

Figure 15:  Foam padding covered in vinyl was placed along the edge of the platform to prevent injury.

Figure 16

Figure 16: This illustration is a close-up view of the padding along the 1" x 4" edge of the mattress platform.

Figure 17

Figure 17: This illustration shows the mattress on its platform. Note the surface of the mattress is nearly level to the highest level of the three-step platform.

Figure 18

Figure 18:  This illustration shows the installation of a standard 30" door and its door hardware.

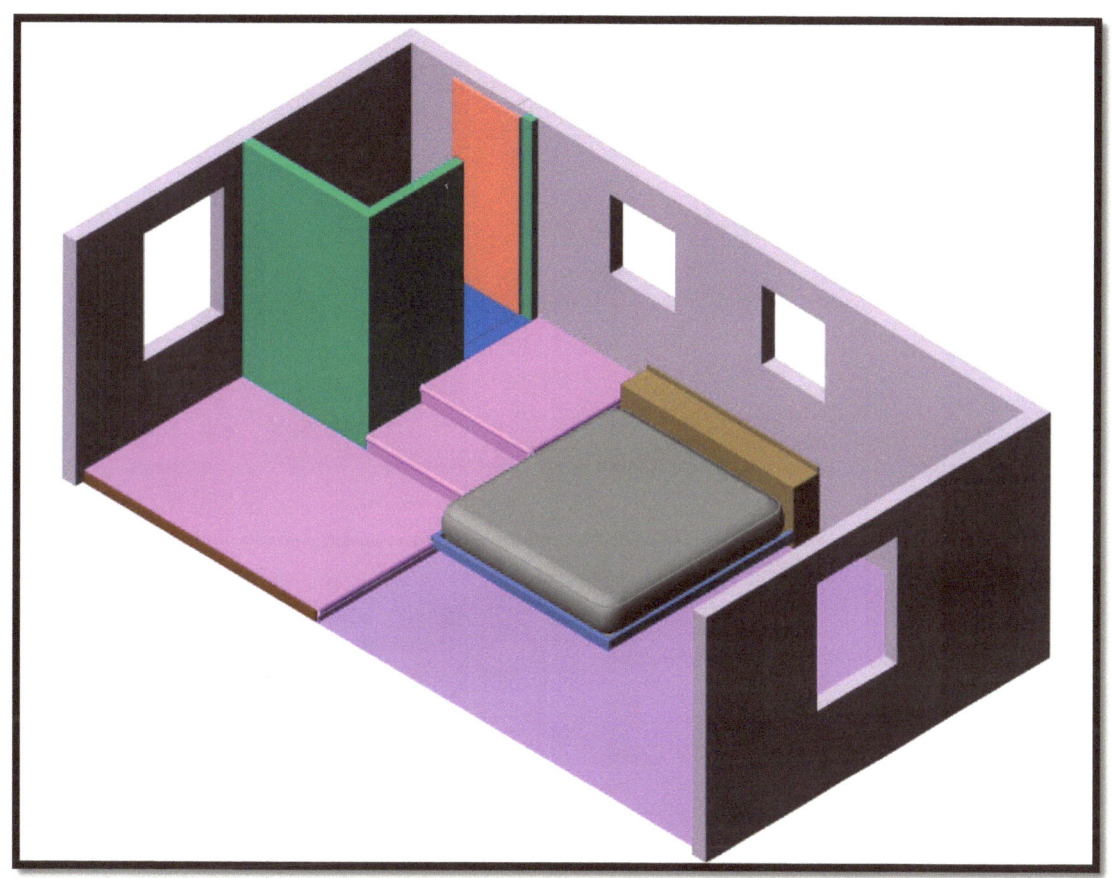

Figure 19

Figure 19:  This illustration shows the completion of the adaptive lavatory, the multi-level platform and the mattress platform.

# Accessible Kitchen Concepts

Let's take a closer look…

Figure 1

Figure 1: I put a cook top stove unit in one corner of the kitchen. Putting the cook top stove unit kitty corner optimizes the surrounding space. I also put an exhaust fan behind the cook top unit so I could put a pot rack above the stove. The pot rack can be repositioned either upward or downward depending on what activities are being performed. There is a crank lever on the side of the appliance cabinet to adjust the pot rack position. Again I avoided putting in any mechanical systems to raise and lower the pot rack that could break down in time. The pot rack has a function and it's aesthetically pleasing.

Figure 2

Figure 2: In this illustration, you can see all the appliances on one wall. I thought it would make more sense to put the microwave and the oven adjacent to the countertop in cases where someone grabs something hotter than they expected and needs to put it down right away, the counter is there for their convenience. If somebody has less hand dexterity, I included an adjustable height portable cart so that the wheelchair user has a very convenient and safe way to get various items from Point A to Point B. The height of the cart can be adjusted by a simple hand crank. In deciding whether to put the microwave above or below the oven, it is fairly certain that in today's fast-paced society, that the microwave gets more use. The bottom of the microwave is 8 inches above the countertop. The oven is directly below the microwave. If a wheelchair user is uncomfortable with the microwave placement, by all means, put it where it is the most convenient to use. The same holds true for the refrigerator. If people are not comfortable with a refrigerator that has the freezer drawer on the bottom, it can always be substituted for a different model. If someone chooses to put one of those compact refrigerator drawer units under the countertop, that might be very smart.

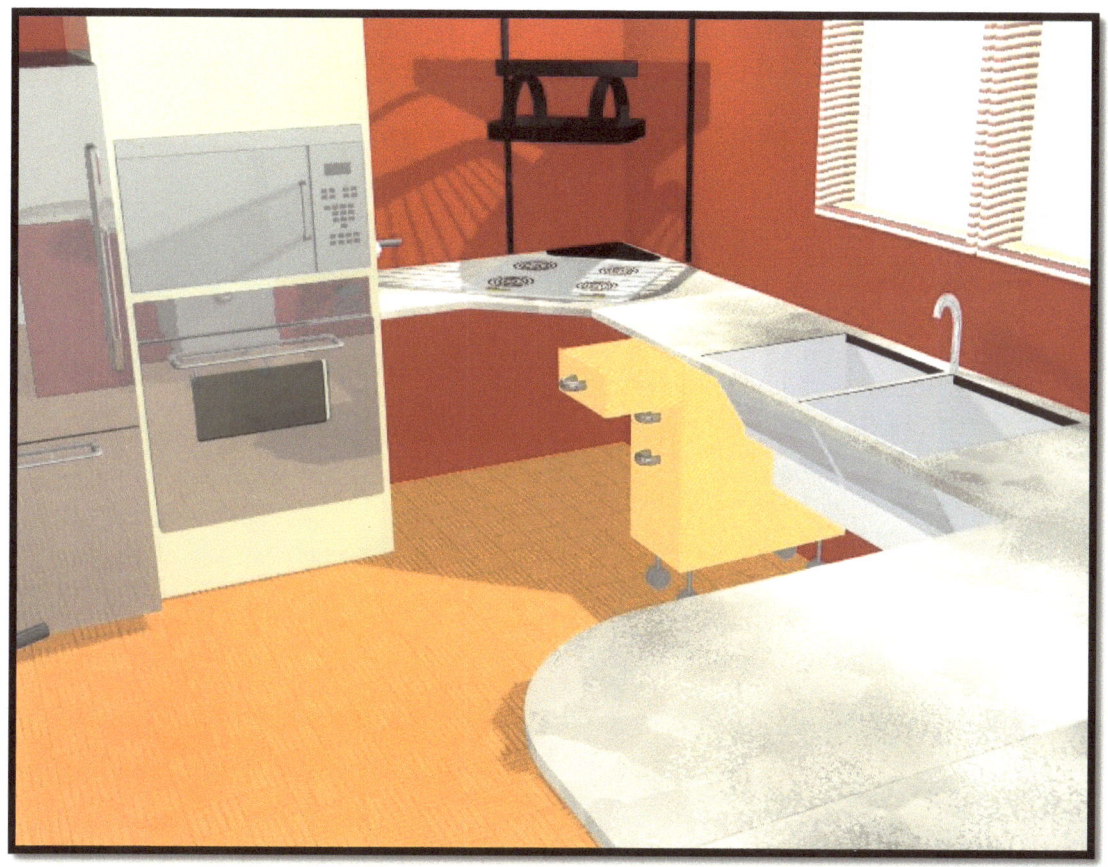

Figure 3

Figure 3: In working with the idea that accessible living spaces can and should be modular, one of my ideas was to put various widths of cabinet drawers on wheels. What is so great about this option is that the wheelchair user can choose how many drawers are needed and where to place them. If someone is right-handed, they can put the set of drawers on the right side of the sink. Left-handed people probably would have the set of drawers on the left side. Underneath the counter would adjustable channels that lock them in place.

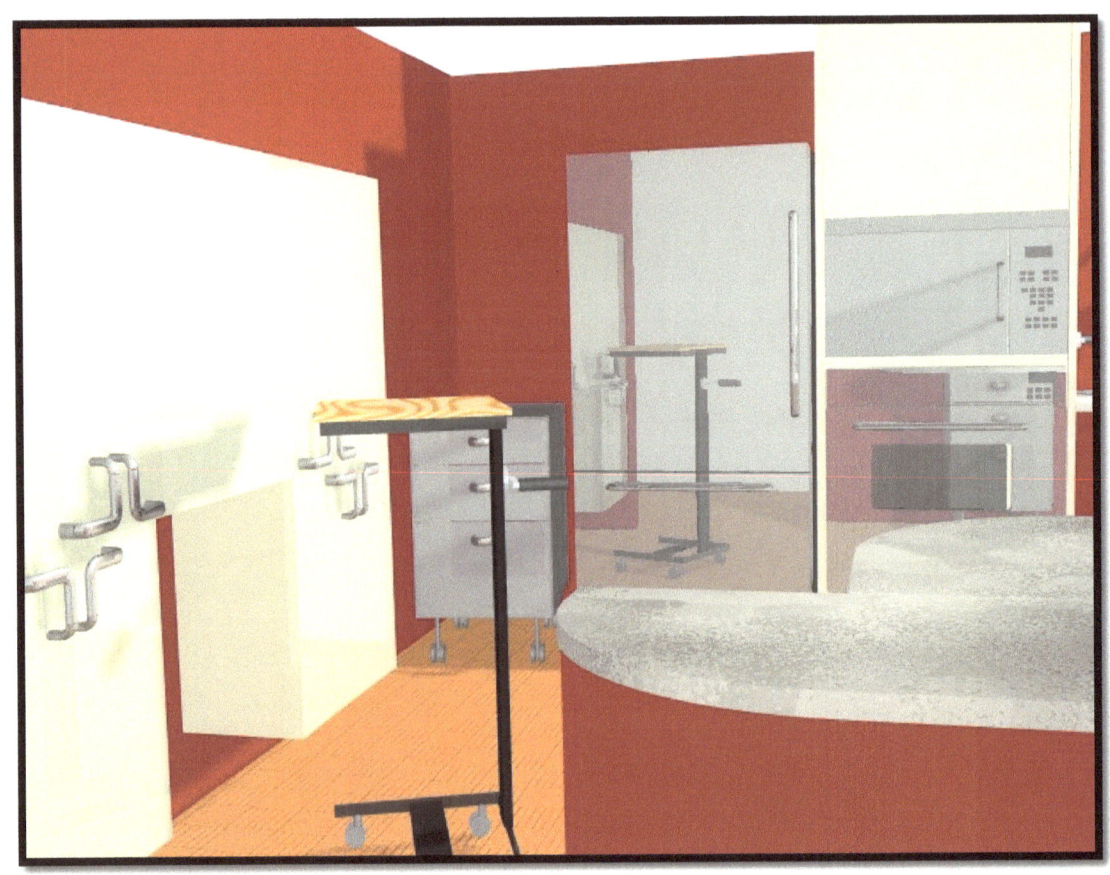

Figure 4

Figure 4:  In this illustration I show another set of drawers tucked away in the corner for future use under the counter, or just for extra storage.  I also show a set of wall cabinets strategically placed for optimal accessibility.  With the wall cabinets only on one wall, it frees up the sight line from outside of the kitchen, into the kitchen.  This is particularly advantageous when you are entertaining a crowd and you want to interact with your guests while still working in the kitchen.  Hanging cabinets won't impede your vision.

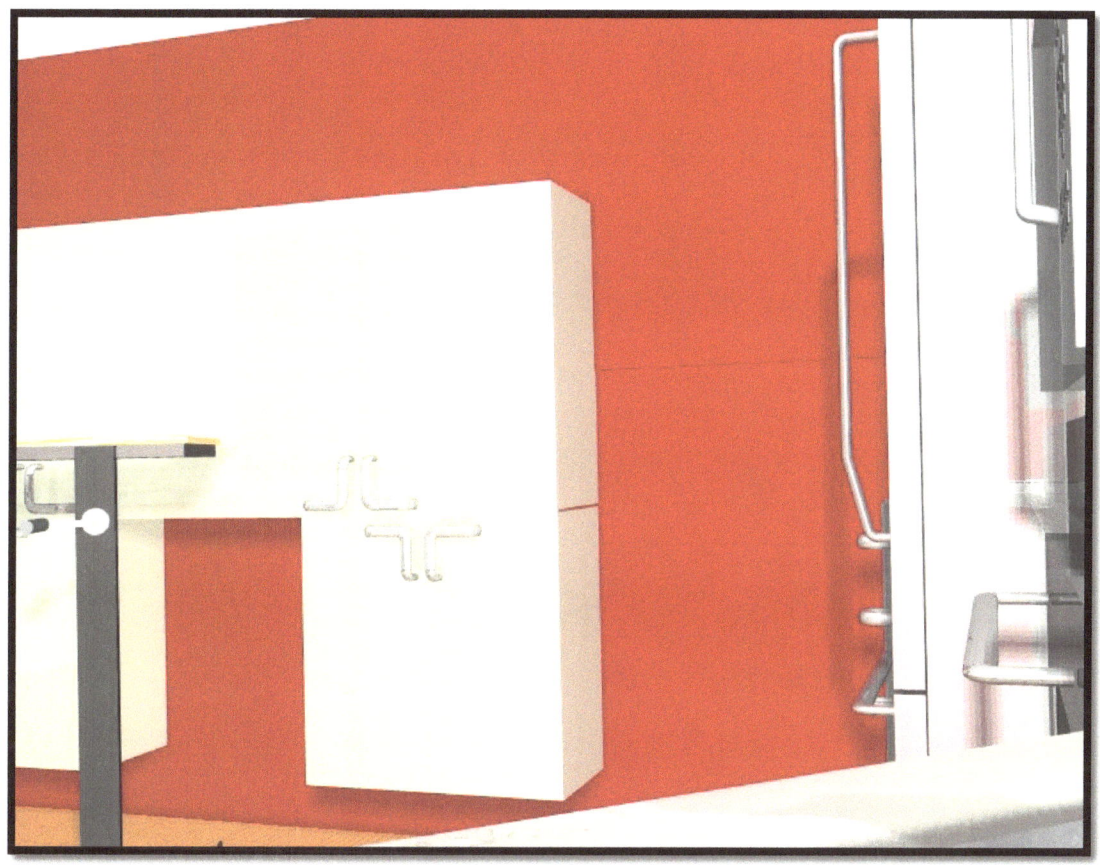

Figure 5

Figure 5:  You can see the bank of wall cabinets situated on the one wall.  There are two sets of 36-inch wide cabinets on top and two sets of 24-inch wide cabinets beneath.  I separated the two lower cabinets to give the wheelchair user an open space to get as close as possible to access the top shelves of the cabinets.  The horizontal line between the upper and lower cabinets is set at 32 inches off the floor.  Here again, I stress the fact that if these wall cabinets are not at a convenient height and or layout, I propose a universal wall cabinet bracket so everything can be changed easily on an individualized basis.  Someone might want more wall cabinets at a lower height.  I think that this system can be utilized with great results.  When the portable cart is not being used, it can fit in between the wall cabinets.

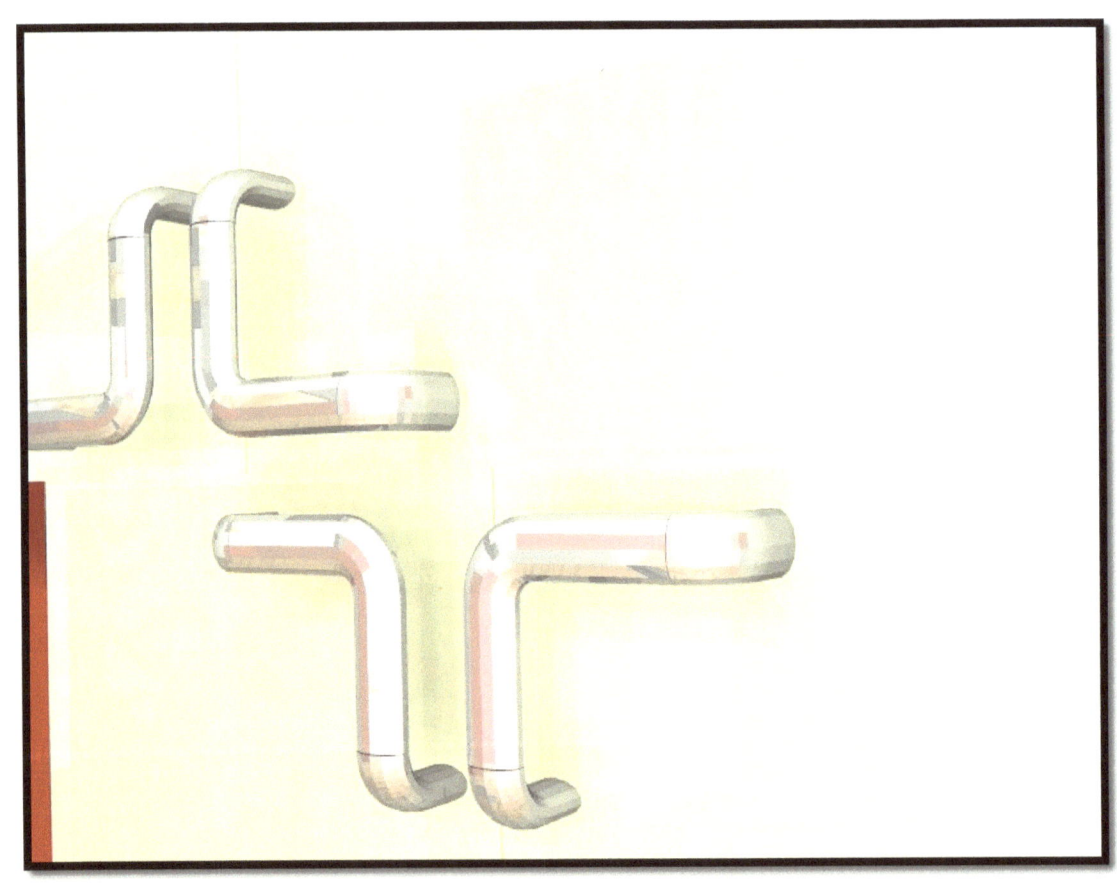

Figure 6

Figure 6: This is a close up view of the cabinet pulls I designed. My thinking behind these was to create a pull that was very functional and highly decorative. I thought about what kind of pull would I be able to grab and operate. I didn't want a pull with any corners that would make it uncomfortable to grab. I didn't want a pull that would be too close to the cabinet door to make it difficult to my fingers around. I came up with this particular pull concept. For the lower wall cabinet door pulls, I flipped the pulls upside down and placed them towards the top of the door.

44

Figure 7

Figure 7:  In staying with the design of the pulls, I came up with an U-shaped drawer pull for people who have minimal dexterity in their hands.  The person would be able to put their hand inside of the loop and pull back on the drawer with minimal effort.  I put the bottom drawer pull towards the top of the drawer so it would be easy to reach.  I would like to come out with an entire line of specialized cabinet and drawer pulls for similar applications.  These pulls can be made out of resin to keep the costs down.

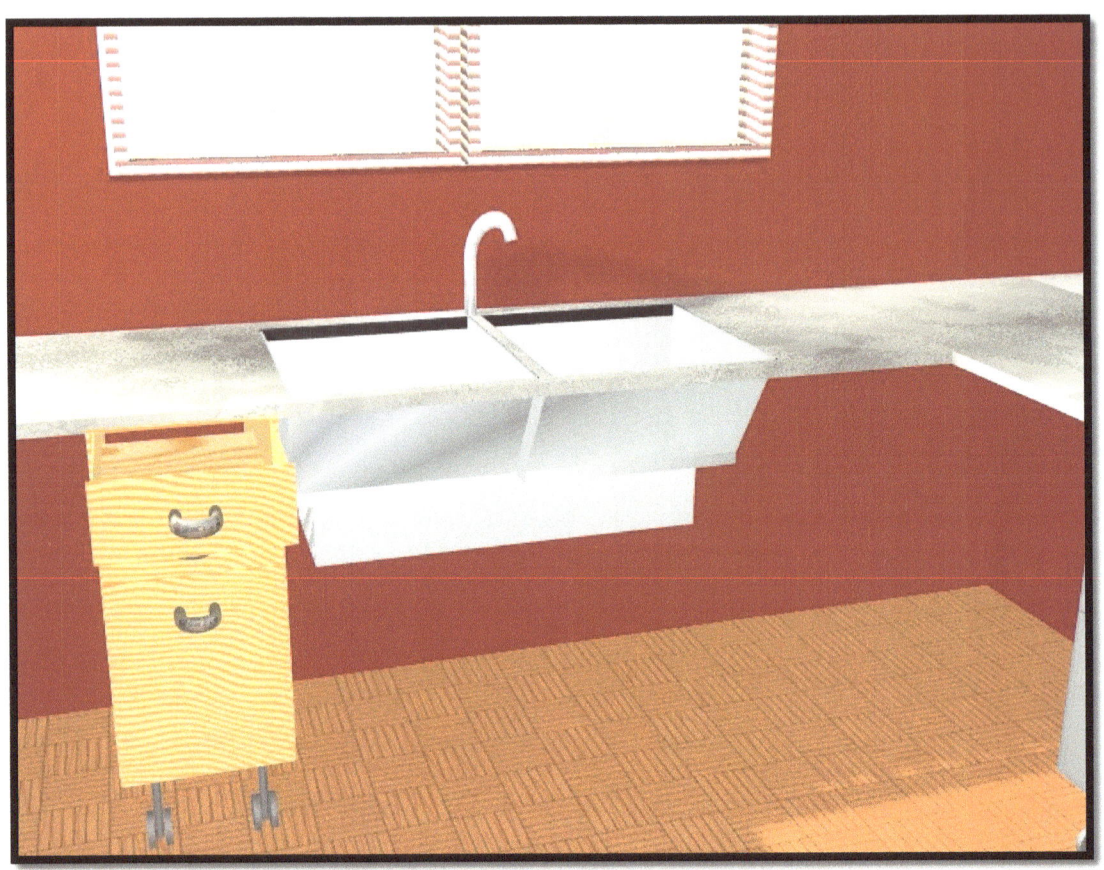

Figure 8

Figure 8:  In this illustration, you can see the countertop and double sinks.  I made the sinks fairly big but not too deep because some people have difficulties reaching down into deep sinks; shoulder rotation is difficult.  Also, with this sink, I slanted in the front side to keep the items in the sink in an easier position to grab.  I have an easier time grabbing things when they are out further away from my body.  When I don't have to control two different actions at once, the less difficulty I have grasping things.  This particular sink is designed with that in mind.  There will be people who work better with things closer to their bodies.  I can design a sink that will suit their needs as well.  It's all about design and how we can form it to make people more independent.  Below the sink, I placed a simple open ended box around the sink plumbing, for the first reason: to protect the wheelchair user from injury, and second: to cover the ugly plumbing.  This box should be easily removable to do maintenance.  I apologize for the poorly drawn sink faucet.  I was working with a new drafting program when I was designing this kitchen, and I didn't want to spend extra time on the faucet when there is a wide selection of sink faucets.  Lastly, on this illustration, I wanted to point out that these cabinets and drawer sets should come in different finishes and veneers.  If anybody is going to live in an apartment or a home for an extended amount of time, they should be able to decorate it as they wish.

# Accessible Bathroom Concepts

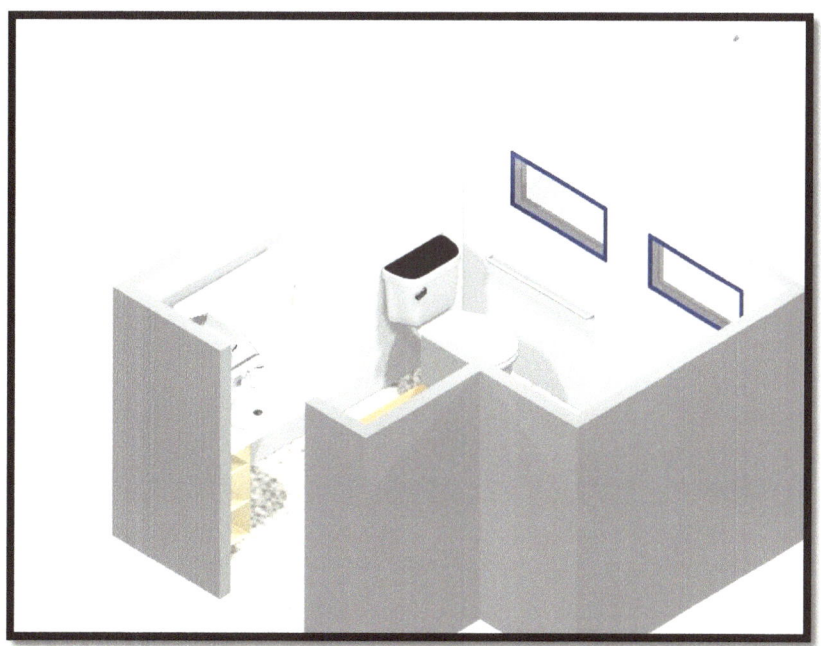

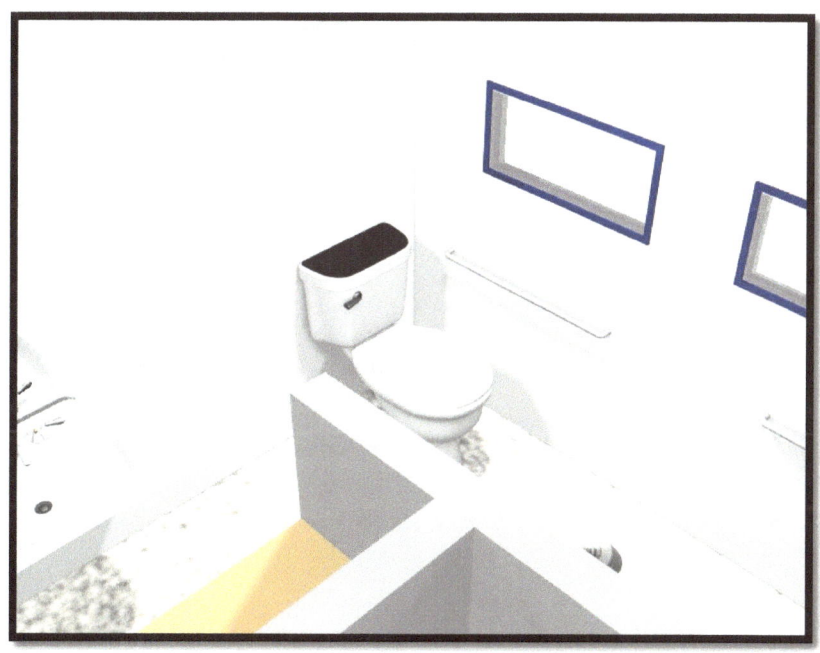

Let's look at some options for assisting with transfers…

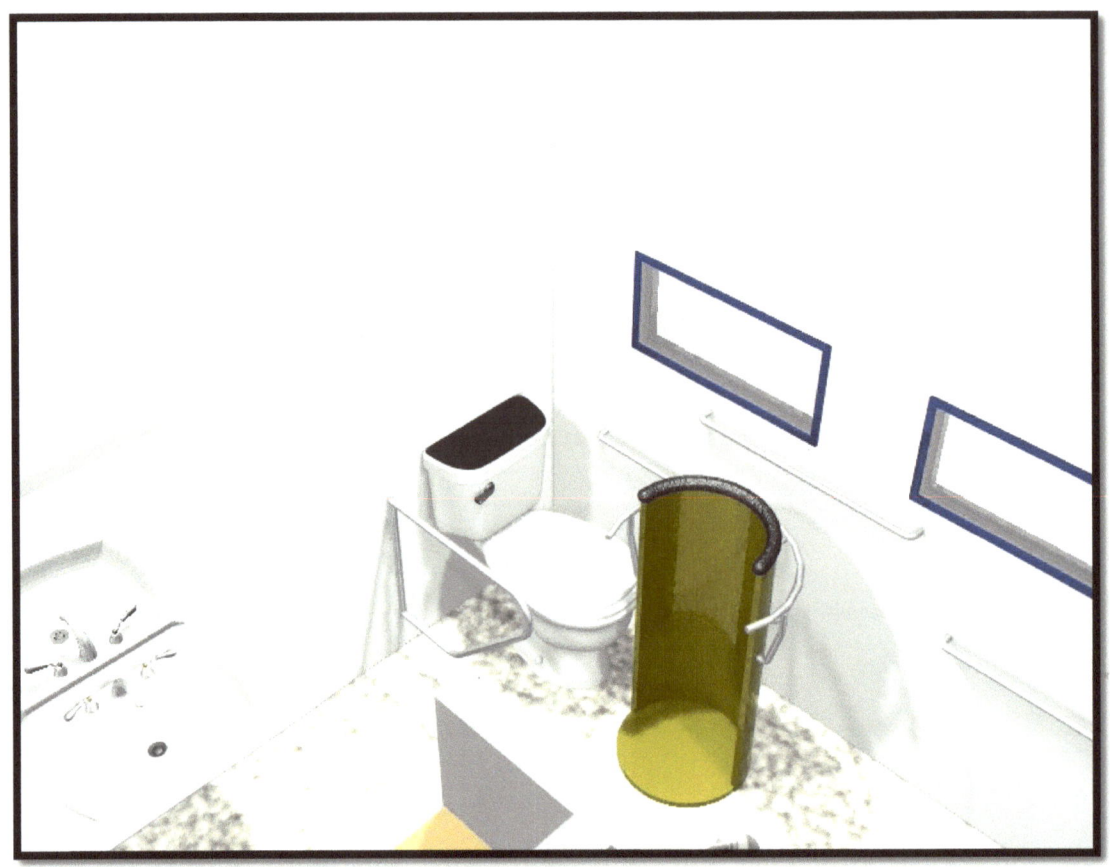

Figure 1

Figure 1: In this particular concept, I devised a half of a cylinder that stands 40 inches in height. For people who have good upper-body strength, but lack lower-body strength and or balance, this pivoting device could be of assistance in becoming more independent. It is sort of a lazy Susan device on a larger scale of course. The top of the cylinder is well padded. There is a circular grab bar around the outer of the cylinder. This circular grab bar gives the person standing a good amount to hold onto. The semi cylinder would be secured to the floor, as its rotation mechanism turns smoothly. The surface of the circular plate would have a non-skid material on it for excellent footing.

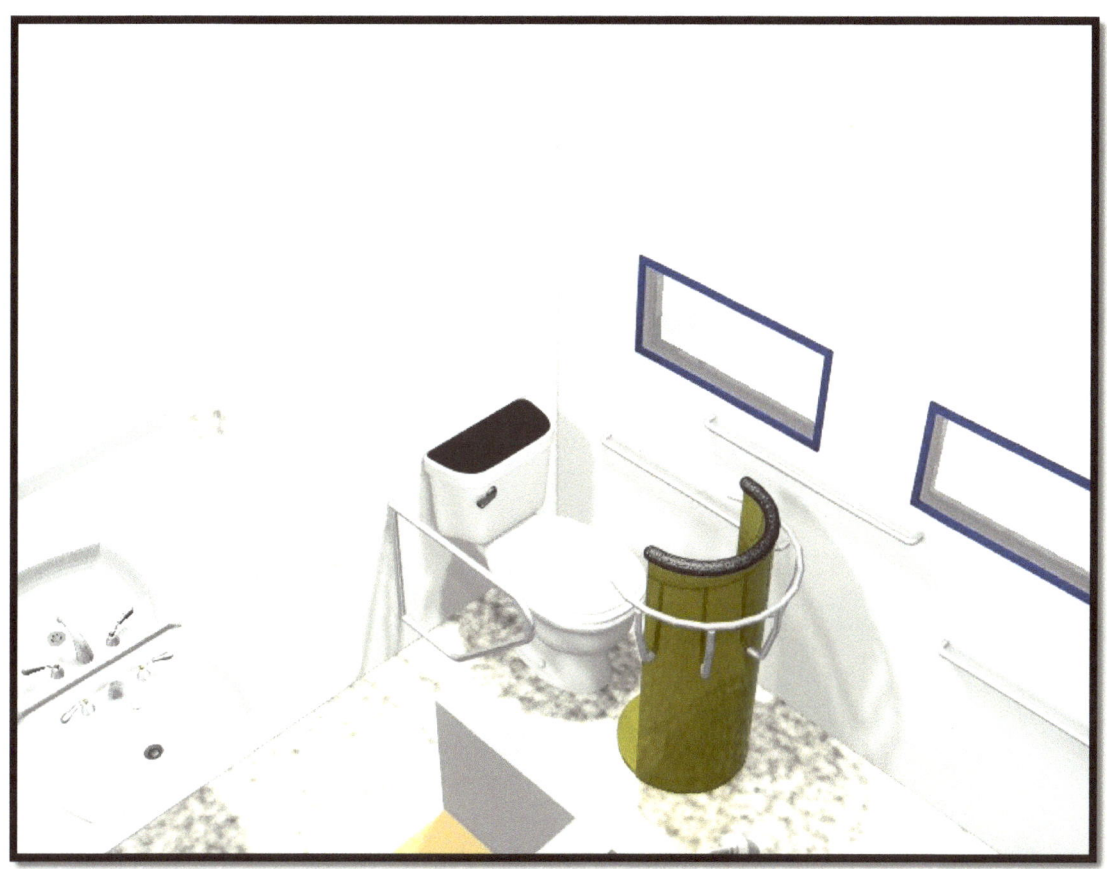

Figure 2

Figure 2: Here, you can see the cylinder rotated 90° to simulate a pivot turn towards the toilet. I placed another grab bar 14 inches higher to assist the person with rotating the cylinder to its desired position.

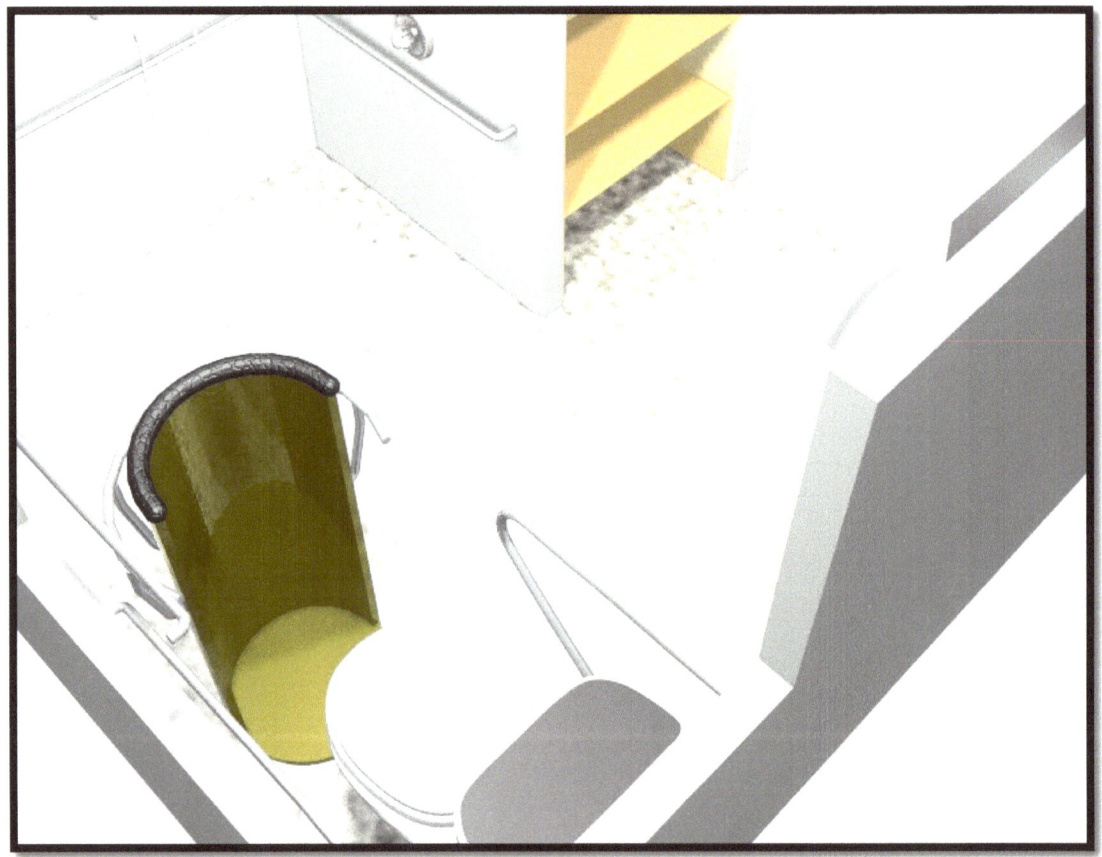

Figure 3

Figure 3: This illustration shows the cylinder at a different angle, from over the toilet. I realize the higher grab bar on the rear wall seems to be levitating in air, but I took away the top of the wall to get this view. I placed a swing away grab bar to the right of the toilet for security.

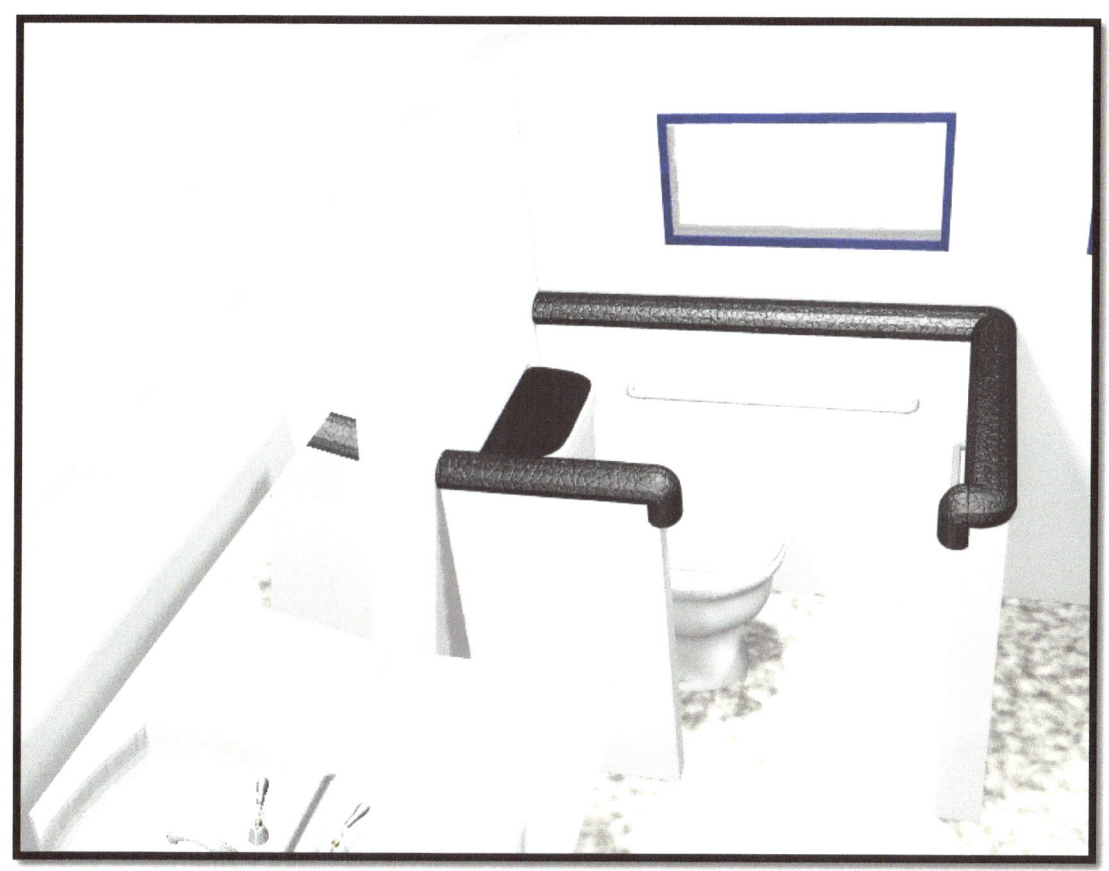

Figure 4

Figure 4: In this second option for assisting people with transfers, I surrounded the toilet with a three, half wall enclosure. It has a space for the wheelchair user to step into. This specific enclosure is more suited for people who have both upper and lower body strength but lack good balance. The enclosure stands 40 inches high and the top is well padded. I placed grab bars all along the inside of the enclosure at 31 inches high, which gives the user solid support. I like this particular enclosure because it totally separate from the shower area and less likely water will invade the toilet enclosure.

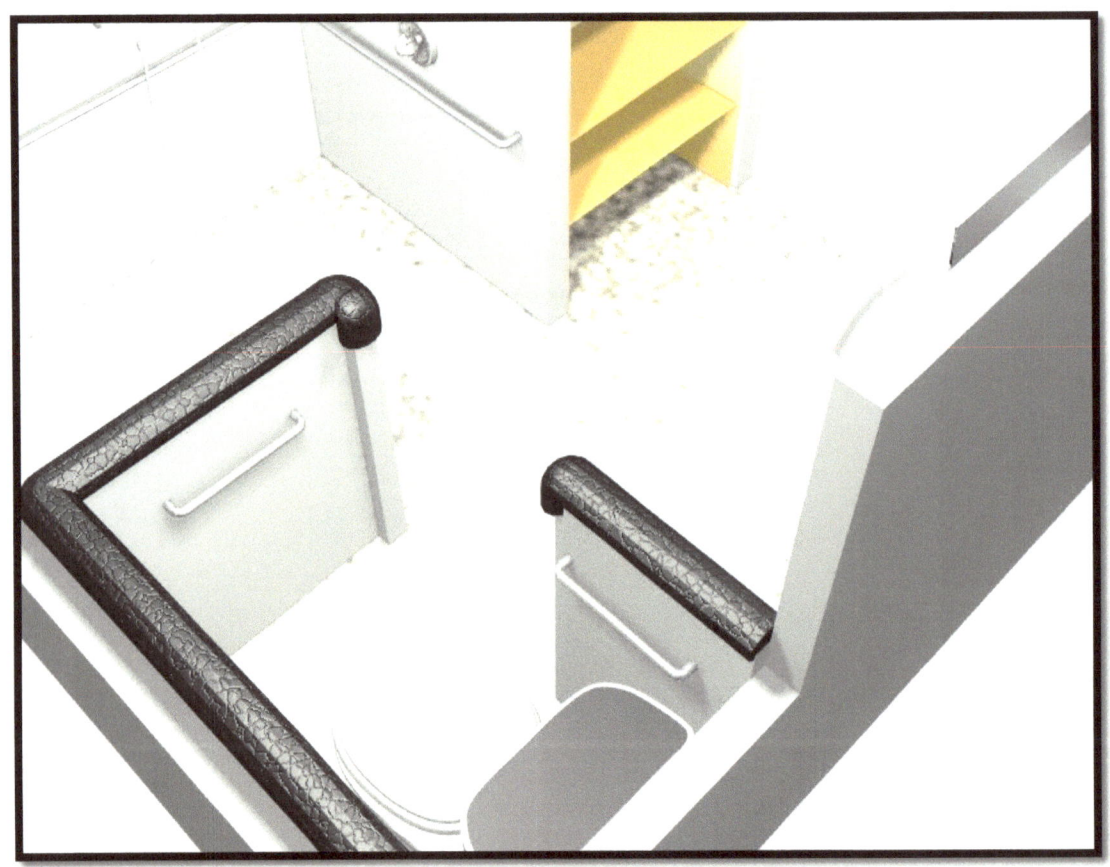

Figure 5

Figure 5: Here is another view of the toilet enclosure taken above the north east corner. You can see how this toilet enclosure can dramatically help keep water in the shower area. The space between the shower wall and the enclosure is 36 inches, so there is plenty of room for a wheelchair or the shower chair to get through.

Maureen Gaynor, using different CAD programs, illustrated all drawings in this book.

www.ingramcontent.com/pod-product-compliance
Lightning Source LLC
Chambersburg PA
CBHW050815180526
45159CB00004B/1673

* 9 7 8 1 5 1 2 1 3 1 6 9 7 *